For 150 years, scientists at the Rothamsted Experimental Station have studied aspects of plant nitrogen nutrition and amino acid biosynthesis. This book is the result of a meeting held to mark this century and a half of work there. The volume considers the significant progress in understanding the biochemistry of amino acids recently achieved in the light of this history of research. Leading researchers from around the world have contributed authoritative chapters on protein amino acids, non-protein amino acids, betaines, glutathione, polyamines, and other secondary metabolites derived from amino acids.

As well as being essential in some animals' nutrition, these compounds can have important roles in defending against herbivores, insects and disease. An understanding of these compounds can help in devising better crop protection and production methods.

T0275878

SOCIETY FOR EXPERIMENTAL BIOLOGY
SEMINAR SERIES: 56

AMINO ACIDS AND THEIR DERIVATIVES IN
HIGHER PLANTS

SOCIETY FOR EXPERIMENTAL BIOLOGY SEMINAR SERIES

A series of multi-author volumes developed from seminars held by the Society for Experimental Biology. Each volume serves not only as an introductory review of a specific topic, but also introduces the reader to experimental evidence to support the theories and principles discussed, and points the way to new research.

AMINO ACIDS AND THEIR DERIVATIVES IN HIGHER PLANTS

Edited by

R.M. Wallsgrove

Biochemistry and Physiology Department,
IACR Rothamsted Experimental Station

CAMBRIDGE
UNIVERSITY PRESS

CAMBRIDGE UNIVERSITY PRESS
Cambridge, New York, Melbourne, Madrid, Cape Town, Singapore, São Paulo

Cambridge University Press
The Edinburgh Building, Cambridge CB2 8RU, UK

Published in the United States of America by Cambridge University Press, New York

www.cambridge.org
Information on this title: www.cambridge.org/9780521454537

First published 1995
This digitally printed version 2008

A catalogue record for this publication is available from the British Library

Library of Congress Cataloguing in Publication data

Amino acids and their derivatives in higher plants/edited by R.M.
Wallsgrove.
   p.   cm.—(Seminar series/Society for Experimental Biology; 53)
   Papers presented at a meeting held at Rothamsted Experimental Station
in September 1993.
   Includes index.
   ISBN 0–521–45453–0
     1. Amino acids—Metabolism—Congresses. 2. Plants—Metabolism—
Congresses. 3. Plant metabolites—Congresses. 4. Botanical chemistry—
Congresses. I. Wallsgrove. R. M. II. Rothamsted Experimental Station.
III. Series: Seminar series (Society for Experimental Biology (Great
Britain)); 53.
QK898.A5A55 1995
582'.019245—dc20 94–31641 CIP

ISBN 978-0-521-45453-7 hardback
ISBN 978-0-521-05051-7 paperback

Contents

Contributors

BENNETT, R.N.
Biochemistry and Physiology Department, Institute of Arable Crops
Research, Rothamsted Experimental Station, Harpenden, Herts AL5
2JQ, UK.

BESFORD, R.T.
Horticulture Research International, Littlehampton, West Sussex
BN17 6LP, UK.

BORRELL, A.
Unitat de Fisiologia Vegetal, Facultat de Farmacia, Universitat de
Barcelona, Av.Diagonal 643, 08028-Barcelona, Spain.

BROWN, E.G.
Biochemistry Research Group, University College of Swansea,
Swansea SA2 8PP, UK.

D'MELLO, J.P.F.
The Scottish Agricultural College, West Mains Road, Edinburgh
EH9 3JG, UK.

DOUCE, R.
Laboratoire de Physiologie Cellulaire Végétale, Centre D'Etudes
Nucléaires de Grenoble et Université Joseph Fourier, CEN-G, 85X,
F-38041, Grenoble Cedex, France.

FORDE, B.G.
Biochemistry and Physiology Department, Institute of Arable Crops
Research, Rothamsted Experimental Station, Harpenden, Herts AL5
2JQ, UK.

FRANKARD, V.
Laboratorium voor Plantengenetica, Institute of Molecular Biology,
Vrije Universiteit Brussel, Paardenstraat 65, B-1640
Sint-Genesius-Rode, Belgium.

GADAL, P.
Physiologie Végétale Moléculaire, URA CNRS 1128, Université de
Paris-Sud, Batiment 430, 91405 Orsay Cedex, France.

GHISLAIN, M.
Laboratorium voor Plantengenetica, Institute of Molecular Biology,
Vrije Universiteit Brussel, Paardenstraat 65, B-1640
Sint-Genesius-Rode, Belgium.
GORHAM, J.
School of Biological Sciences, University of Wales, Bangor,
Gwynedd, Wales, UK.
HALKIER, B.A.
Plant Biochemistry Laboratory, Department of Plant Biology, The
Royal Veterinary and Agricultural University, Thorvaldsensvej 40,
DK-1781 Frederiksberg C, Copenhagen, Denmark.
HILTZ, D.A.
Biology Department, Mount Allison University, Sackville, New
Brunswick E0A 3C0, Canada.
IRELAND, R.J.
Biology Department, Mount Allison University, Sackville, New
Brunswick E0A 3C0, Canada.
JACOBS, M.
Laboratorium voor Plantengenetica, Institute of Molecular Biology,
Vrije Universiteit Brussel, Paardenstraat 65, B-1640
Sint-Genesius-Rode, Belgium
JOHN, P.
Department of Agricultural Botany, University of Reading,
Whiteknights PO Box 221, Reading RG6 2AS, UK.
KOCH, B.M.
Plant Biochemistry Laboratory, Department of Plant Biology, The
Royal Veterinary and Agricultural University, Thorvaldsensvej 40,
DK-1781 Frederiksberg C, Copenhagen, Denmark.
LEPINIEC, L.
Physiologie Végétale Moléculaire, URA CNRS 1128, Université de
Paris-Sud, Batiment 430, 91405 Orsay Cedex, France.
MAHER, S.E.
Institute of Sustainable Irrigated Agriculture, Department of
Agriculture, Tatura, Vic. 3616, Australia.
MARCÉ, M.
Unitat de Fisiologia Vegetal, Facultat de Farmacia, Universitat de
Barcelona, Av.Diagonal 643, 08028-Barcelona, Spain.
MITHEN, R.
Brassica and Oilseeds Research Department, John Innes Centre,
Institute for Plant Science Research, Colney Lane, Norwich NR4
7UJ, UK.

MØLLER, B.L.
Plant Biochemistry Laboratory, Department of Plant Biology, The
Royal Veterinary and Agricultural University, Thorvaldsensvej 40,
DK-1781 Frederiksberg C, Copenhagen, Denmark.
NARDELLA, N.E.
Institute of Sustainable Irrigated Agriculture, Department of
Agriculture, Tatura, Vic. 3616, Australia.
OLIVER, D.
Department of Molecular Biology and Biochemistry, University of
Idaho, Moscow, ID 83843, USA.
RAWSTHORNE, S.
Brassica and Oilseeds Research Department, John Innes Centre,
Colney, Norwich, NR4 7UJ, UK.
RENNENBERG, H.
Institute of Forest Botany & Tree Physiology, University of
Freiburg, Am Flughafen 17, D-79085 Freiburg i. Br., Germany.
ROUZÉ, P.
Plant Biochemistry Laboratory, Department of Plant Biology, The
Royal Veterinary and Agricultural University, Thorvaldsensvej 40,
DK-1781 Frederiksberg C, Copenhagen, Denmark.
SANTI, S.
Istituto di Produzione Vegetale, Universita degli Studi Di Udine,
Via Fagagna 208, 33 100 Udine, Italy.
SHANER, D.L.
American Cyanamid Company, PO Box 400, Princeton, NJ
08543-0400, USA.
SIBBESEN, O.
Plant Biochemistry Laboratory, Department of Plant Biology, The
Royal Veterinary and Agricultural University, Thorvaldsensvej 40,
DK-1781 Frederiksberg C, Copenhagen, Denmark.
SINGH, B.K.
American Cyanamid Company, PO Box 400, Princeton, NJ
08543-0400, USA.
SZAMOSI, I.
American Cyanamid Company, PO Box 400, Princeton, NJ
08543-0400, USA.
TIBURCIO, A.F.
Unitat de Fisiologia Vegetal, Facultat de Farmacia, Universitat de
Barcelona, Av.Diagonal 643, 08028-Barcelona, Spain.
TOROSER, D.
Brassica and Oilseeds Research Department, John Innes Centre,

Institute for Plant Science Research, Colney Lane, Norwich, NR4
7UJ, UK.
UGALDE, T.D.
Institute of Sustainable Irrigated Agriculture, Department of
Agriculture, Tatura, Vic. 3616, Australia.
VAUTERIN, M.
Laboratorium voor Plantengenetica, Institute of Molecular Biology,
Vrije Universiteit Brussel, Paardenstraat 65, B-1640
Sint-Genesius-Rode, Belgium.
WALLSGROVE, R.M.
Biochemistry and Physiology Department, Institute of Arable Crops
Research, Rothamsted Experimental Station, Harpenden, Herts AL5
2JQ, UK.
WOODALL, J.
Biochemistry and Physiology Department, Institute of Arable Crops
Research, Rothamsted Experimental Station, Harpenden, Herts AL5
2JQ, UK.

Preface

This book arose from a meeting at Rothamsted Experimental Station in September 1993, sponsored by the Plant Metabolism Group of the SEB. It was one of several meetings, on a wide variety of topics, which helped to celebrate the 150th anniversary of Rothamsted, which was founded by John Bennett Lawes in 1843. At that time, Lawes was involved in the commercial production of superphosphate fertilizer, and the experiments begun by Lawes and his co-worker Gilbert investigated many aspects of plant nutrition. That the conference (and this book) dealt with amino acids, end products of nitrogen assimilation in plants, is most appropriate considering the pioneering work of Lawes and Gilbert on plant nitrogen nutrition.

Amino acid biochemistry in plants has been a major topic of research at Rothamsted for more than 20 years, and many major advances have been made: the discovery of the glutamate synthase cycle for the assimilation of ammonia, the first description of the photorespiratory nitrogen cycle, and detailed genetic, biochemical and molecular analysis of these processes; the biochemistry and genetics of amino acid biosynthesis, particularly aspartate-derived amino acids; and most recently the biochemistry of amino acid-derived secondary metabolites. This work has involved many other researchers from laboratories all over the world, and it was a great pleasure to welcome so many past and present collaborators to the meeting, and to read of their latest work in the chapters of these proceedings. One aspect that was very clear at the meeting, and I hope is reflected in the book, is the need to integrate many disciplines if we are to understand the biology of amino acids in plants: physiology, genetics, biochemistry and molecular biology all have a part to play, and none of them alone can provide all the answers. The fastest and most secure progress will come from integrated approaches covering all these disciplines and methodologies.

Many other topics are covered in this volume, yet even so we have not been able to cover the whole field of amino acid metabolism in plants – no single volume could! In particular, the amazing variety of non-protein amino acids and amino acid-derived secondary metabolites are only superficially covered, and we have not included any chapter on the shikimic acid pathway and the vast range of metabolites derived from it. Fortunately, aromatic amino acid biosynthesis and metabolism have been comprehensively reviewed elsewhere. Several other books have recently dealt with aspects of amino acids in plants, and it is a measure of the scale of the field that there is relatively little duplication or overlap with the current volume!

There are still many aspects of amino acid metabolism in plants that are only poorly understood. The biosynthesis, and metabolic regulation of synthesis, of several protein amino acids still pose questions, such as histidine synthesis, the regulation of methionine synthesis, (neither covered here), and proline metabolism, which is dealt with in a limited way in this volume. Amino acid-derived secondary metabolites are still a rich field for biochemists, and recent progress in understanding cyanogenic glucoside and glucosinolate metabolism is encouraging. The differences between these superficially similar pathways are intriguing, and suggest that the plant kingdom still has many surprises for us. The meeting at Rothamsted has, I hope, encouraged new interactions and stimulated current ones between workers in the field, to take research forward in many of these (and other) areas.

I would like to record my appreciation of the financial and other assistance provided by Rothamsted, Ciba-Geigy, Schering AG, Cyanamid, and Unilever, plus the encouragement and assistance provided by Steve Rawsthorne of the Plant Metabolism Group. Many people helped to make the meeting a success, and to produce this book, and my thanks go to all of my colleagues for their assistance and support.

Roger Wallsgrove
Rothamsted, March 1994

BRIAN G. FORDE and JANET WOODALL

Glutamine synthetase in higher plants: molecular biology meets plant physiology

Glutamine synthetase (GS; EC 6.3.1.2) is a key enzyme of plant N metabolism, occurring in most or all plant tissues and contributing to a wide variety of physiological processes. This chapter looks at some recent developments in the molecular biology of the enzyme, and how these are leading to new insights into the participation of individual isoenzymes in different metabolic pathways.

Glutamine synthetase catalyses the incorporation of ammonium into glutamine, using glutamate as substrate. In higher plants it is the major enzyme responsible for the assimilation of ammonium (Miflin & Lea, 1980), acting in conjunction with glutamate synthase (GOGAT) to synthesize both glutamine and glutamate, the precursors of all other organic nitrogenous compounds in the plant. Not only is GS involved in primary N assimilation (using ammonium absorbed from the soil or generated by either NO_3^- reduction or symbiotic N fixation), but also in the reassimilation of the N released endogenously by a variety of ammonium-evolving processes, such as photorespiration, phenylpropanoid metabolism and the transamination of amino acids (Fig. 1; for review see Joy, 1988).

The complexity of the role that GS plays in plant N metabolism is compounded by the spatial and temporal diversity of the pathways of primary and secondary N assimilation. Some ammonium-generating processes (for example, N fixation in legumes) are restricted to a single cell type (infected cells of root or stem nodules). Others, such as photorespiration, are temporally regulated or, in the case of NO_3^- or ammonium assimilation in roots, are subject to unpredictable fluctuations in the external N source.

Given the above considerations, it is perhaps not surprising that molecular studies are revealing a complex system for regulating the biosynthesis of GS in higher plants. It now appears that, in most plant species, GS is encoded by a multigene family consisting of a minimum of four functional genes which encode one plastidic and at least three

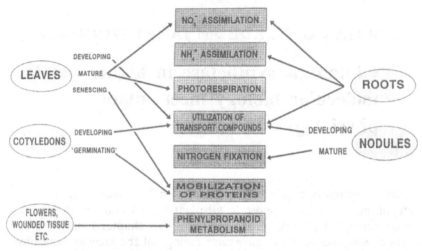

Fig. 1. Metabolic pathways involving GS in higher plants and their occurrence in different parts of the plant at different stages of development.

cytosolic GS polypeptides. This holds true in pea (Tingey, Walker & Coruzzi, 1987; Tingey *et al.*, 1988), French bean (Cullimore *et al.*, 1984; Gebhardt *et al.*, 1986; Lightfoot, Green & Cullimore, 1988), soybean (Miao *et al.*, 1991; Roche, Temple & Sengupta-Gopalan, 1993), Arabidopsis (Peterman & Goodman, 1991) and maize (Sakakibara *et al.*, 1992*b*; Li *et al.*, 1993). A possible exception is mustard, which may only have a single (plastidic) GS gene (Höpfner, Ochs & Wild, 1991). Initial studies indicated that, apart from the induction of the chloroplast enzyme by light (Lightfoot *et al.*, 1988; Tingey *et al.*, 1988), expression of individual GS genes was primarily under developmental control, with each organ displaying a different complement of GS mRNAs, polypeptides and isoenzymes (Forde & Cullimore, 1989). Since 1989, however, it has become evident that this is an over-simplification. In this review we will examine recent developments relating to the spatial distribution of GS isoenzymes in plant organs and the nutritional regulation of GS gene expression. In the final section we will discuss the prospects for genetic manipulation of GS in higher plants and review current progress in this area.

Spatial distribution of GS isoenzymes in plant tissues

Root nodules

When the promoters of two cytosolic GS genes from French bean (*gln*-β and *gln*-γ) were fused to the GUS reporter gene (*uidA*) and introduced into *Lotus corniculatus*, it was possible to use a histochemical GUS assay to compare the spatial expression patterns of the two genes (Forde *et al.*, 1989). These studies confirmed that the nodule-enhanced expression of *gln*-γ and the root-enhanced expression of *gln*-β, which had been established from mRNA studies in French bean (Gebhardt *et al.*, 1986; Bennett, Lightfoot & Cullimore, 1989), are conferred by the transcriptional properties of their promoters. However, more significantly, the histochemical analysis showed that although both genes are expressed in nodules, they display quite different spatial patterns of expression. Thus *gln*-γ expression was localized to the central infected zone of the nodule and was strongest in the infected cells themselves. Expression of *gln*-β, on the other hand, occurred throughout the nodule in the early stages of nodulation, but in mature nodules its expression became restricted to the vascular strands.

Subsequent cytological studies using *in situ* hybridization and gene-specific probes (Teverson, 1990), looked at the distribution of GS mRNAs in French bean nodules at different stages of development. Confirming the results obtained with transgenic *Lotus*, it was found that the *gln*-γ mRNA was abundant in infected cells, but significant amounts could also be detected in the non-infected cells of the central zone and in the inner cortex. The *gln*-β mRNA was present throughout the nodule in the early stages of nodulation and its abundance rapidly declined with nodule maturity, particularly in the infected cells. In the mature and late nodule, *gln*-β mRNA was mainly confined to the cortex (inner, mid- and outer) and particularly to the vascular endodermis. Consistent with the distribution of the mRNAs, Chen & Cullimore (1989) found when they dissected French bean nodules, that GS in the cortex was mainly composed of the β subunit, while in the central tissue the γ subunit predominated.

The spatial separation of the two GS subunits within the nodule (although not complete) explains why there are two distinct cytosolic isoenzymes in French bean nodules, one (GS_{n1}) nodule-specific and composed of γ with some β, and the other (GS_{n2}) closely related to the root form and almost entirely a homooctamer of β (Lara *et al.*, 1984; Bennett & Cullimore, 1989). Thus GS_{n1} and GS_{n2}, which differ in their kinetic properties and their ratios of transferase to synthetase activities (Cullimore *et al.*, 1983), are also located in different regions of the nodule.

What does the spatial distribution of GS_{n1} and GS_{n2} tell us about their likely metabolic roles? The localization of the nodule-specific GS_{n1} isoenzyme in the central infected zone of the nodule is consistent with it having the major role in the assimilation of the ammonium released by the N-fixing bacteroids. However, the role of GS_{n2} is less clear.

One possibility is that it too is involved in primary ammonium assimilation, despite its physical separation from the infected cells. This could happen if excess ammonium, unassimilated by GS_{n1}, diffuses into the cortical region. Perhaps more likely, given its association with the nodule's vascular system, is that GS_{n2} has a different role that is in some way related to transport of nitrogenous compounds. If it is not assimilating the products of primary N assimilation, it must be assimilating ammonium released from other N-containing compounds. Possibilities include the deamination of amino acids or the catabolism of ureides. In alfalfa and *L. corniculatus*, ^{14}C-labelling studies have found that while little radiolabelled glutamine was present in nodules, glutamine was the major labelled amino acid in root extracts and was the third most abundant amino acid in xylem sap (Maxwell *et al.*, 1984). These observations suggest that there is significant recycling of nitrogenous compounds during their movement through the xylem to the shoot, involving lateral abstraction of N solutes from the xylem sap and reloading of the xylem with glutamine and other amino acids. Although the function of this recycling is unclear, it is possible that it begins even in the nodule vascular system and that the vascular-associated GS_{n2} could play a key role in the process.

An alternative source of amino acids for recycling through GS_{n2} is the phloem sap supplying root nodules. Although the primary function of phloem transport to the nodule is to provide carbon for growth, respiration and amino acid biosynthesis, the phloem sap also contains significant amounts of N (Pate *et al.*, 1979; Layzell *et al.*, 1981; Jeschke, Atkins & Pate, 1984). For white lupin and soybean, it has been estimated that as much as 15% of the N exported through the xylem to the shoot is returned to the nodule via the phloem (Walsh, 1990; Parsons *et al.*, 1993). The returning amino acids and amides may well undergo some transformations in the nodule while in transit between phloem and xylem, and again this is likely to require GS activity in the vicinity of the vascular tissue.

Parsons *et al.* (1993) have proposed a model for the regulation of nodule growth and activity in which the concentration of reduced N compounds flowing into the nodule from the phloem is used as a signal indicating the nitrogen status of the shoot. It is suggested that changes in phloem N content, in rates of C and oxygen supply, or in amide or ureide accumulation, are integrated into an effect on the concentration of a key compound (such as asparagine or glutamine) in the nodule cortex. In the model, the concentration of this key compound then regulates the short-term and long-term oxygen diffusion barriers. It may also have a feedback effect on N fixation and N assimilation.

Such a model may explain the unexpected finding that growth and N fixation are markedly stimulated in alfalfa plants in which root GS and the root form of GS in nodules (equivalent to GS_{n2}) are inhibited by a bacterial toxin (tabtoxin), that has no effect on the nodule-specific isoform (Knight & Langston-Unkefer, 1988). If, as in French bean, the root isoform of GS in nodules is associated with the vascular tissue, its inhibition would be expected to have a severe effect on N cycling, perhaps reducing the concentration of the key N compound and signalling a low N status in the shoot.

Other plant organs

The GUS reporter gene was also used to compare the expression patterns of the pea plastidic (GS2) and cytosolic (GS3A) GS genes in transgenic tobacco plants (Edwards, Walker & Coruzzi, 1990). It was found that the GS2 promoter conferred light-regulated expression on the GUS gene in photosynthetic cells of leaves, stems and cotyledons. In contrast, the GS3A promoter directed expression specifically in the phloem of leaves, stems and roots of mature plants, and vascular-specificity was also seen in the cotyledons, hypocotyl and roots of germinating seedlings. Similar results were obtained for the GS3A promoter in transgenic alfalfa plants (Brears, Walker & Coruzzi, 1991).

The GS_2 promoter from French bean has been shown to direct expression in palisade and mesophyll parenchyma of transgenic tobacco leaves, and, at much lower levels, in the leaf epidermis, but not in the pith parenchyma or the vascular tissue (Cock, Hémon & Cullimore, 1992). Three recent studies using immunocytological methods have confirmed the vascular localization of GS_1 and the mesophyll localization of GS_2 in leaves of rice (Kamachi *et al.*, 1992), potato (Pereira *et al.*, 1992) and tobacco (Carvalho *et al.*, 1992). In potato and tobacco the cytosolic GS protein was confined to the phloem companion cells.

There is therefore an accumulating body of evidence suggesting that the cytosolic and plastidic GS isoenzymes are spatially separated in the leaves of many plant species. However, these findings are at variance with earlier results obtained with protoplasts isolated from leaves of barley and pea, which indicated that a significant proportion of GS activity in mesophyll cells (from 15 to 50%) is in the cytosolic fraction (Wallsgrove, Lea & Miflin, 1979; Wallsgrove *et al.*, 1980). This discrepancy will need to be resolved by further experiments.

The physiological implications of a differential distribution of GS isoenzymes within the leaf were discussed by Edwards *et al.* (1990). These authors concluded that the spatial separation of GS_1 and GS_2

indicated they had non-overlapping roles: GS_2 being primarily involved in assimilating ammonium released in photosynthetic cells by photorespiration and nitrate/nitrite reduction, and GS_1 in the phloem companion cells having a role in intercellular glutamine transport.

Regulation of GS by N nutrition

A number of recent studies have addressed the question of whether GS expression in plants is regulated by either ammonium or NO_3^-.

Ammonium

The ability of externally applied ammonium to induce the expression of a soybean GS gene in roots has been demonstrated (Hirel et al., 1987; Miao et al., 1991). The promoter of the soybean GS15 gene, which encodes a cytosolic GS polypeptide in roots and nodules, conferred ammonium-inducible expression on the GUS gene in transgenic L. corniculatus plants: expression was increased more than three-fold by 10 mM ammonium sulphate, while application of KNO_3, asparagine or glutamine had no significant effect (Miao et al., 1991). Earlier studies had shown that the ammonium effect could be seen in soybean roots at the mRNA level within 2 h of treatment (Hirel et al., 1987). Progressive 5' deletion of the GS15 promoter located the ammonium-responsive elements to the region between 3.5 and 1.3 kb upstream of the transcription start site (Marsolier, Carrayol & Hirel, 1993).

However it is clear that not all GS genes are ammonium inducible. There are many instances in the literature where ammonium treatments had little or no effect on GS activity (for example, Mann, Fentem & Stewart, 1980; Loyola-Vargas & Sánchéz de Jiménez, 1986; Vézina & Langlois, 1989; Cock et al., 1990; Shen, 1991). In both pea and French bean, when photorespiratory ammonium production was suppressed by growth at increased CO_2 concentrations, there was a long-term effect that resulted in reduced levels of GS_2 mRNA (but not GS_1 mRNA) in leaves (Edwards & Coruzzi, 1989; Cock et al., 1991). However, as this effect was not seen in short-term experiments (Cock et al., 1991), it is likely to be due to an indirect effect on leaf metabolite concentrations that develops on long-term exposure to high CO_2, rather than to a direct effect of the absence of photorespiratory ammonium production.

There can be numerous reasons for the conflicting evidence concerning the ammonium inducibility of GS genes in plants. One of these is certainly that not all GS genes possess the ammonium-responsive cis-acting elements found in the soybean GS15 gene. However there are other important factors which relate to the experimental protocols that are used to assess ammonium inducibility. The first of these arises from the use of alternative assay methods to monitor GS induction, that is mRNA, protein or enzyme activity. For example, in French bean, Hoelzle *et al.* (1992) found that a 28 h ammonium treatment led to a five-fold induction of root GS_1 activity, but in agreement with the mRNA studies of Cock *et al.* (1990) this occurred without any increase in GS_1 protein. These results emphasize the importance of following gene expression at all levels, since it seems that changes in GS activity can occur without changes in GS protein or mRNA, and it is very likely that the converse is also true.

The second important consideration is our lack of knowledge about the nature of the plant's response to ammonium treatment. There are many factors (genetic, physiological, developmental and environmental) that will determine how adding ammonium will affect the cellular ammonium concentration and the distribution of ammonium between the vacuole and the cytosol (Lee & Ratcliffe, 1991). Because the cytosolic fraction makes only a small contribution to the cellular ammonium concentration, measurements of tissue ammonium content really reflect the vacuolar concentration (Lee & Ratcliffe, 1991). The rate of assimilation of ammonium by GS has been shown to be an important factor in determining the extent to which ammonium accumulates in the cytoplasm (Lee & Ratcliffe, 1991), so that cells with a low basal level of GS activity are more likely than those with high basal levels of GS activity to experience a surge in cytoplasmic ammonium after ammonium is added to the external medium. We cannot even be certain that it is ammonium itself that is directly responsible for the induction of GS genes when this occurs: it is quite possible that the response is to fluctuations in other metabolites that are affected by the ammonium treatment. Thus, when the GS genes in a particular species or tissue are found not to be inducible by ammonium, this does not necessarily imply the lack of 'ammonium-responsive' *cis*-acting elements, but may reflect the nature of the physiological response to ammonium in that species/tissue. This conclusion is supported by the finding that the ammonium-inducible soybean GS15 promoter is unresponsive to ammonium treatment in transgenic tobacco, even though the same construct functions as expected in *L. corniculatus* (Miao *et al.*, 1991).

Nitrate

Nitrate is well known to be able to elicit the rapid induction (<2 h) of nitrate reductase (NR) and nitrite reductase (NiR), a response that is regulated at the transcriptional level (Pelsy & Caboche, 1992). The first report that GS activity was also subject to NO_3^- regulation came from pea roots, where a 2.5-fold increase in plastidic GS activity was observed after 8 h (Emes & Fowler, 1983). Subsequently it was shown that this induction was specific to the plastidic form of the enzyme, occurred at the protein level and was specific to NO_3^- (Vézina & Langlois, 1989). Similar results have been obtained with *L. corniculatus*, as shown in Fig. 2. A NO_3^--specific induction of GS_2 mRNA and protein in maize roots has also been reported (Sakakibara *et al.*, 1992*a*; Redinbaugh & Campbell, 1993) and has been shown to involve an

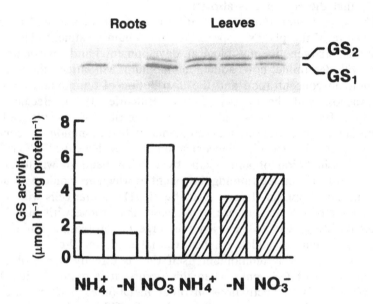

Fig. 2. Nitrate-inducibility of GS_2 in roots of *L. corniculatus*. Plants were grown under sterile conditions for 47 d in culture on N-free medium or medium containing ammonium (5 mM) or NO_3^- (5 mM) as indicated. GS activity was assayed by the semi-biosynthetic method and extracts of roots or leaves containing 2 μg soluble protein were electrophoresed and western blotted. GS protein was detected using rabbit antiserum raised against nodule GS from French bean (Cullimore & Miflin, 1984) and an anti-rabbit ExtrAvidin-Peroxidase kit (Sigma Chemical Co. Ltd).

accumulation of the GS_2 mRNA and to be accompanied by an induction of mRNA for ferredoxin-dependent glutamate synthase (Fd-GOGAT) (Redinbaugh & Campbell, 1993). Two proteins that provide reducing power for nitrite reduction and Fd-GOGAT activity, ferredoxin-$NADP^+$ oxidoreductase and ferredoxin, are also induced by NO_3^- in pea root plastids (Bowsher, Hucklesby & Emes, 1993).

Analysis of the kinetics of mRNA induction in maize roots showed that induction of the GS_2 gene occurs coordinately (within 30 min of treatment) with that of the NR gene and that, like the induction of NR and NiR, it is insensitive to cycloheximide and is thus part of the root's primary response to environmental NO_3^- (Redinbaugh & Campbell, 1991, 1993). In contrast, in maize leaves, the GS_2 and Fd-GOGAT genes are expressed constitutively and are not subject to NO_3^- inducibility (Redinbaugh & Campbell, 1993).

These observations imply a specific role for GS_2 in some species in the assimilation of ammonium generated endogenously by the NO_3^- assimilatory pathway in roots, but not in the assimilation of ammonium taken up directly from the soil. The latter function must be adequately covered by the cytosolic isoenzyme(s). A corollary of this hypothesis is that there should be a distinction between species that assimilate NO_3^- in their roots and those that are shoot assimilators in the occurrence of plastidic GS in their roots. It has been proposed that, in general, temperate legume species growing in low external NO_3^- concentrations are primarily root assimilators, while tropical and sub-tropical species are primarily shoot assimilators (Andrews, 1986a).

In a survey of 43 species of the Papilionoideae (grown in 1 mM NO_3^-; J.W. and B.G.F, unpublished observations), a GS_2-like polypeptide was found in the roots of each of 30 temperate species (representing 26 genera and encompassing 15 tribes), but was undetectable in the roots of any of 17 tropical or sub-tropical species (14 genera and 6 tribes) (Fig. 3). This strong correlation amongst such a phylogenetically diverse group of legumes suggests that, in the papilionoid legumes at least, the possession of a root plastidic GS isoenzyme is part of some form of adaptation to a temperate climate. Since many of the temperate legume species are primarily root assimilators of NO_3^-, it is likely that this isoform of GS does play a role in the NO_3^- assimilatory pathway. However, based on *in vivo* and *in vitro* NR assays, many of the temperate legumes resembled the tropical and sub-tropical legumes in having more than 50% of their NR activity in the shoot. Although NR assays are not a definitive method for determining the proportion of nitrate reduced in the root (Andrews, 1986b), it is possible that the occurrence of plastidic GS in the root is a more distinctive feature of temperate legumes than is root NO_3^- assimilation.

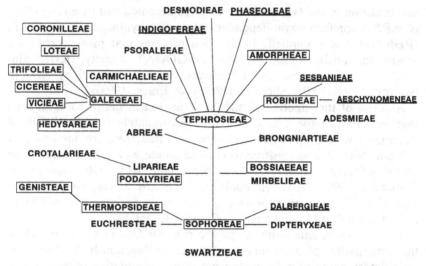

Fig. 3. Distribution amongst the tribes of the Papilionoideae of a GS_2-like polypeptide in roots. The phylogenetic tree is taken from Corby, Polhill & Sprent (1983). Boxes indicate those tribes in which all species tested by western blotting (see Fig. 2) have a GS_2-like polypeptide in the roots when grown on NO_3^-. Tribes in which no species with a root GS_2 have been found are underlined. In all cases, temperate species had a GS_2-like polypeptide in their roots, while tropical and sub-tropical species did not. The Tephrosieae (circled) include both temperate and tropical species, and when one of each type was tested the temperate species (*Wisteria floribunda*) had a GS_2 polypeptide, while the tropical species (*Tephrosia vogelii*) did not.

Genetic manipulation of glutamine synthetase

The key role that GS plays in plant N metabolism, the multiplicity of GS isoenzymes and the diversity of pathways in which they are potentially involved, combine to make GS a particularly attractive target for genetic manipulation. The identification of GS_2-deficient mutants of barley was important in establishing the role of this isoenzyme in the photorespiratory nitrogen cycle (Blackwell, Murray & Lea, 1987; Wallsgrove *et al.*, 1987), but selection methods that would allow the isolation of other types of GS mutant have not been devised. The availability of additional mutants that are deficient in a specific GS isoenzyme, or that have a particular organ or cell type that is deficient in that isoform, would be invaluable in establishing the role(s) that individual isoenzymes play in N metabolism, and their contribution to metabolic flux through ammonium assimilatory pathways.

The development of antisense technology for down-regulating plant gene expression (Schuch, 1991) holds out the prospect of generating such 'mutant' lines by combining developmentally regulated promoters with gene-specific antisense constructs. To answer some of the questions raised earlier in this review, one could envisage generating transgenic legumes that are GS_2-deficient specifically in the roots, or in which GS_1 had been down-regulated in the phloem or in which the root GS isoform was eliminated from the nodule. [Intriguingly, there is evidence for a natural antisense mechanism for regulating GS gene expression in the bacterium *Clostridium acetobutylicum* (Fierro-Monti, Reid & Woods, 1992)].

In the only report so far of the use of antisense to manipulate GS in a transgenic plant, an alfalfa GS_1 cDNA was fused in the antisense orientation to a constitutive promoter (the 35S promoter from the cauliflower mosaic virus) and the construct introduced into tobacco (Temple *et al.*, 1993). Two of the antisense lines were found to have 25–40% reduced levels of GS activity per g fresh weight of leaves, and the same lines accumulated about 30% less soluble protein per g fresh weight. As the GS activity was reduced to about the same degree as the general pool of soluble proteins, it was not clear whether this was a direct effect of the antisense gene or part of a more general effect on plant performance.

To date, most attempts to manipulate GS activity in transgenic plants have concentrated on overexpression of the enzyme. These studies have all involved fusion of the CaMV 35S promoter to a GS_1 coding sequence (from either a cDNA or genomic clone) and transformation of tobacco (*Nicotiana tabacum*) or *N. plumbaginifolia* (Eckes *et al.*, 1989; Hémon, Robbins & Cullimore, 1990; Hirel *et al.*, 1992; Temple *et al.*, 1993) and are reviewed in more detail elsewhere (Lea & Forde, 1994). The first such experiments had the objective of generating plants that were resistant to phosphinothricin, a herbicide that is a specific inhibitor of GS, and resulted in a five-fold increase in GS activity in the transgenic tobacco leaves (Eckes *et al.*, 1989). On the assumption that most of this novel GS_1 activity is likely to be in the leaf mesophyll cells, which in wild-type tobacco lack significant cytosolic GS activity (McNally *et al.*, 1983; Hirel *et al.*, 1992), this represents a major change in the compartmentation of GS in the transformed plants. Amino acid analysis showed that the overproduction of GS had no effect on the pool of free amino acids, but did result in a seven-fold reduction in free ammonium levels. The latter effect suggests that GS activity in the tobacco leaf is limiting the rate of ammonium assimilation, but it is unclear whether it is the increased overall GS activity or the novel presence of GS activity in the cytosolic compartment of the leaf mesophyll cell that is important. The other attempts to overexpress GS (Hémon *et al.*, 1990; Hirel *et al.*, 1992; Temple *et al.*, 1993) have been unable to reproduce the high levels of GS activity obtained by Eckes and colleagues and have not shed any additional light on the physiological consequences of GS overproduction.

Clearly, the potential for increased understanding of the partitioning and regulation of glutamine biosynthetic pathways that is offered by genetic manipulation has yet to be fully realised. A further dimension is added to the prospects for manipulating GS by the possible role of glutamine as a key regulatory molecule in higher plants as it is in bacteria (Stock, Stock & Mottonen, 1990). In plants, glutamine (and/or a metabolite derived directly from it) has been implicated in the feedback repression of NO_3^- and ammonium absorption by maize roots (Lee *et al.*, 1992) and of NR gene expression in *Nicotiana* spp. (Deng *et al.*, 1991; Vincentz *et al.*, 1993) and in the induction of phosphoenolpyruvate carboxylase and other enzymes of C4 metabolism in maize (Sugiharto *et al.*, 1992; Sugiharto & Sugiyama, 1992). Thus alterations in GS (or Fd-GOGAT) activity that lead to changes in the glutamine pool may have indirect consequences for both N and C metabolism that exceed those expected from simply modulating the flux through the ammonium assimilatory pathway. As discussed above, a regulatory role for glutamine may help to explain why a treatment that inhibits 'root' GS in roots and root nodules of alfalfa stimulates N fixation and nodulation (Knight & Langston-Unkefer, 1989) and why leaf N content increases in oat plants that have had almost all their root GS activity inhibited by the same treatment (Knight, Bush & Langston-Unkefer, 1988).

References

Andrews, M. (1986*a*). The partitioning of nitrate assimilation between root and shoot of higher plants. *Plant, Cell and Environment* **9**, 511–19.

Andrews, M. (1986*b*). Nitrate and reduced-N concentrations in the xylem sap of *Stellaria media*, *Xanthium strumarium* and six legume species. *Plant, Cell and Environment*, **9**, 605–8.

Bennett, M.J. & Cullimore, J.V. (1989). Glutamine synthetase isoenzymes of *Phaseolus vulgaris* L.: subunit composition in developing root nodules and plumules. *Planta*, **179**, 433–40.

Bennett, M.J., Lightfoot, D.A. & Cullimore, J.V. (1989). cDNA sequence and differential expression of the gene encoding the glutamine synthetase γ polypeptide of *Phaseolus vulgaris* L. *Plant Molecular Biology,* **12**, 553–65.

Blackwell, R.D., Murray, A.J.S. & Lea, P.J. (1987). Inhibition of photosynthesis in barley with decreased levels of chloroplastic glutamine synthetase activity. *Journal of Experimental Botany,* **38**, 1799–809.

Bowsher, C.G., Hucklesby, D.P. & Emes, M.J. (1993). Induction of ferredoxin-NADP⁺ oxidoreductase and ferredoxin synthesis in pea root plastids during nitrate assimilation. *The Plant Journal,* **3**, 463–7.

Brears, T., Walker, E.L. & Coruzzi, G.M. (1991). A promoter sequence involved in cell-specific expression of the pea glutamine synthetase GS3A gene in organs of transgenic tobacco and alfalfa. *The Plant Journal,* **1**, 235–44.

Carvalho, H., Pereira, S., Sunkel, C. & Salema, R. (1992). Detection of a cytosolic glutamine synthetase in leaves of *Nicotiana tabacum* L. by immunocytochemical methods. *Plant Physiology,* **100**, 1591–4.

Chen, F-L. & Cullimore, J.V. (1989). Location of two isoenzymes of NADH-dependent glutamate synthase in root nodules of *Phaseolus vulgaris* L. *Planta,* **179**, 441–7.

Cock, J.M., Brock, I.W., Watson, A.T., Swarup, R., Morby, A.P. & Cullimore, J.V. (1991). Regulation of glutamine synthetase genes in leaves of *Phaseolus vulgaris*. *Plant Molecular Biology,* **17**, 761–71.

Cock, J.M., Hémon, P. & Cullimore, J.V. (1992). Characterization of the gene encoding the plastid-located glutamine synthetase of *Phaseolus vulgaris*: regulation of β-glucuronidase gene fusions in transgenic tobacco. *Plant Molecular Biology,* **18**, 1141–9.

Cock, J.M., Mould, R.M., Bennett, M.J. & Cullimore, J.V. (1990). Expression of glutamine synthetase genes in roots and nodules of *Phaseolus vulgaris* following changes in the ammonium supply and infection with various *Rhizobium* mutants. *Plant Molecular Biology,* **14**, 549–60.

Corby, H.D.L., Polhill, R.M. & Sprent, J.I. (1983). Taxonomy. In *Nitrogen Fixation, Vol. 3: Legumes*, ed. W.J. Broughton, pp 1–35. Oxford: Oxford University Press.

Cullimore, J.V., Gebhardt, C., Saarelainen, R., Miflin, B.J., Idler, K.B. & Barker, R.F. (1984). Glutamine synthetase of *Phaseolus vulgaris* L.: organ-specific expression of a multigene family. *Journal of Molecular and Applied Genetics* **2**, 589–99.

Cullimore, J.V., Lara, M., Lea, P.J. & Miflin, B.J. (1983). Purification and properties of two forms of glutamine synthetase from the plant fraction of *Phaseolus* root nodules. *Planta,* **157**, 245–53.

Cullimore, J.V. & Miflin, B.J. (1984). Immunological studies on glutamine synthetase using antisera raised against two plant forms of the enzyme from *Phaseolus* root nodules. *Journal of Experimental Botany,* **35**, 581–7.

Deng, M-D., Moureaux, T., Cherel, I., Boutin, J-P. & Caboche, M. (1991). Effects of nitrogen metabolites on the regulation and circadian expression of tobacco nitrate reductase. *Plant Physiology and Biochemistry*, **29**, 239–47.

Eckes, P., Schmitt, P., Daub, W. & Wengenmayer, F. (1989). Overproduction of alfalfa glutamine synthetase in transgenic plants. *Molecular and General Genetics*, **217**, 263–8.

Edwards, J.W. & Coruzzi, G.M. (1989). Photorespiration and light act in concert to regulate the expression of the nuclear gene for chloroplast glutamine synthetase. *The Plant Cell*, **1**, 241–8.

Edwards, J.W., Walker, E.L. & Coruzzi, G.M. (1990). Cell-specific expression intransgenic plants reveals non-overlapping roles for chloroplast and cytosolic glutamine synthetase. *Proceedings of the National Academy of Sciences, USA*, **87**, 3459–63.

Emes, M.J. & Fowler, M.W. (1983). The supply of reducing power for nitrite reduction in plastids of seedling pea roots (*Pisum sativum* L.). *Planta*, **158**, 97–102.

Fierro-Monti, I.P., Reid, S.J. & Woods, D.R. (1992). Differential expression of a *Clostridium acetobutylicum* antisense RNA: implications for regulation of glutamine synthetase. *Journal of Bacteriology*, **174**, 7642–7.

Forde, B.G. & Cullimore, J.V. (1989). The molecular biology of glutamine synthetase in higher plants. *Oxford Surveys of Plant Molecular and Cell Biology*, **6**, 247–96.

Forde, B.G., Day, H.M., Turton, J.F., Shen, W.-J., Cullimore, J.V. & Oliver, J.E.(1989). Two glutamine synthetase genes from *Phaseolus vulgaris* L. display contrasting developmental and spatial patterns of expression in transgenic *Lotus corniculatus* plants. *Plant Cell*, **1**, 391–401.

Gebhardt, C., Oliver, J.E., Forde, B.G., Saarelainen, R. & Miflin, B.J. (1986). Primary structure and differential expression of glutamine synthetase genes in nodules, roots and leaves of *Phaseolus vulgaris*. *The EMBO Journal*, **5**, 1429–35.

Hémon, P., Robbins, M.P. and Cullimore, J.V. (1990). Targeting of glutamine synthetase to the mitochondria of transgenic tobacco. *Plant Molecular Biology*, **15**, 895–904.

Hirel, B., Bouet, C., King, B., Layzell, D., Jacobs, F. & Verma, D.P. (1987). Glutamine synthetase genes are regulated by ammonia provided externally or by symbiotic nitrogen fixation. *The EMBO Journal*, **6**, 1167–71.

Hirel, B., Marsolier, M.C., Hoarau, A., Hoarau, J., Brangeon, J., Schafer, R. &Verma, D.P.S. (1992). Forcing expression of a soybean glutamine synthetase gene in tobacco leaves induces a native gene encoding cytosolic enzyme. *Plant Molecular Biology*, **20**, 207–18.

Hoelzle, I., Finer, J.J., McMullen, M.D. & Streeter, J.G. (1992). Induction of glutamine synthetase activity in non-nodulated roots of *Glycine max, Phaseolus vulgaris* and *Pisum sativum. Plant Physiology,* **100**, 525–8.

Höpfner, M., Ochs, G. & Wild, A. (1991). Detection of a single gene encoding glutamine synthetase in *Sinapis alba* (L.). *Journal of Plant Physiology,* **139**, 76–81.

Jeschke, W.D., Atkins, C.A. & Pate, J.S. (1984). Ion circulation via phloem and xylem between the root and shoot of nodulated white clover. *Journal of Plant Physiology,* **117**, 319–30.

Joy, K.W. (1988). Ammonia, glutamine, and asparagine: a carbon–nitrogen interface. *Canadian Journal of Botany,* **66**, 2103–9.

Kamachi, K., Yamaya, T., Hayakawa, T., Mae, T. & Ojima, K. (1992). Vascular bundle-specific localization of cytosolic glutamine synthetase in rice leaves. *Plant Physiology,* **99**, 1481–6.

Knight, T.J. & Langston-Unkefer, P.J. (1988). Enhancement of symbiotic dinitrogen fixation by a toxin-releasing plant pathogen. *Science,* **241**, 951–4.

Knight, T.J., Bush, D.R. & Langston-Unkefer, P.J. (1988). Oats tolerant of *Pseudomonas syringae* pv tabaci contain tabtoxin-beta-lactam-insensitive leaf glutamine synthetases. *Plant Physiology,* **88**, 333–9.

Lara, M., Cullimore, J.V., Lea, P.J., Miflin, B.J., Johnston, A.W.B. & Lamb, J.W. (1984). Appearance of a novel form of glutamine synthetase during nodule development in *Phaseolus vulgaris* L. *Planta,* **157**, 254–8.

Layzell, D.B., Pate, J.S., Atkins, C.A. & Canvin, D.T. (1981). Partitioning of carbon and nitrogen and the nutrition of root and shoot apex in a nodulated legume. *Plant Physiology,* **67**, 30–6.

Lea, P.J. & Forde, B.G. (1994). The use of mutants and transgenic plants to study amino acid metabolism. *Plant Cell and Environment,* **17**, 541–56.

Lee, R.B., Purves, J.V., Ratcliffe, R.G. & Saker, L.R. (1992). Nitrogen assimilation and the control of ammonium and nitrate absorption by maize roots. *Journal of Experimental Botany,* **43**, 1385–96.

Lee, R.B. & Ratcliffe, R.G. (1991). Observations on the subcellular distribution of the ammonium ion in maize root tissue using *in vivo* ^{14}N-nuclear magnetic resonance spectroscopy. *Planta,* **183**, 359–67.

Li, M.-G., Villemur, R., Hussey, P.J., Silflow, C.D., Gantt, J.S. & Snustad, D.P.(1993). Differential expression of six glutamine synthetase genes in *Zea mays. Plant Molecular Biology,* **23**, 401–7.

Lightfoot, D.A., Green, N.K. & Cullimore, J.V. (1988). The chloroplast-located glutamine synthetase of *Phaseolus vulgaris* L.: nucleotide sequence, expression in different organs and uptake into isolated chloroplasts. *Plant Molecular Biology,* **11**, 191–202.

Loyola-Vargas, V.M. & Sánchez de Jiménez, E. (1986). Regulation of glutamine synthetase/glutamate synthase cycle in maize tissues. Effect of the nitrogen source. *Journal of Plant Physiology,* **124**, 147–54.

McNally, S.F., Hirel, B., Gadal, P., Mann, A.F. & Stewart, G.R. (1983). Glutamine synthetases of higher plants: evidence for a specific isoform content related to their possible physiological role and their compartmentation within the leaf. *Plant Physiology*, **72**, 22–5.

Mann, A.F., Fentem, P.A. & Stewart, G.R. (1980). Tissue localization of barley (*Hordeum vulgare*) glutamine synthetase isoenzymes. *FEBS Letters*, **110**, 265–7.

Marsolier, M-C., Carrayol, E. & Hirel, B. (1993). Multiple functions of promoter sequences involved in organ-specific expression and ammonia regulation of a cytosolic soybean glutamine synthetase gene in transgenic *Lotus corniculatus*. *The Plant Journal*, **3**, 405–14.

Maxwell, C.A., Vance, C.P., Heichel, G.H. & Stade, S. (1984). CO_2 fixation in alfalfa and birdsfoot trefoil root nodules and partitioning of ^{14}C to the plant. *Crop Science*, **24**, 257–64.

Miao, G-H., Hirel, B., Marsolier, M.C., Ridge, R.W. & Verma, D.P.S. (1991). Ammonia-regulated expression of a soybean gene encoding cytosolic glutamine synthetase in transgenic *Lotus corniculatus*. *The Plant Cell*, **3**, 11–22.

Miflin, B.J. & Lea, P.J. (1980). Ammonia assimilation. In *The Biochemistry of Plants, Vol. 5*, ed. B.J. Miflin, pp. 169–202. New York: Academic Press.

Parsons, R., Stanforth, A., Raven, J.A. & Sprent, J.I. (1993). Nodule growth and activity may be regulated by a feedback mechanism involving phloem nitrogen. *Plant, Cell and Environment* **16**, 125–36.

Pate, J.S., Atkins, C.A., Hamel, K., McNeil, D.L. & Layzell, D.B. (1979). Transport of organic solutes in phloem and xylem of a nodulated legume. *Plant Physiology*, **63**, 1082–8.

Pelsy, F. and Caboche, M. (1992). Molecular genetics of nitrate reductase in higher plants. *Advances in Genetics*, **30**, 1–40.

Pereira, S., Carvalho, H., Sunkel, C. & Salema, R. (1992). Immunocytolocalization of glutamine synthetase in mesophyll and phloem of leaves of *Solanum tuberosum* L. *Protoplasma*, **167**, 66–73.

Peterman, T.K. & Goodman, H.M. (1991). The glutamine synthetase gene family of *Arabidopsis thaliana*: light-regulation and differential expression in leaves, roots and seeds. *Molecular and General Genetics*, **230**, 145–54.

Redinbaugh, M.G. & Campbell, W.H. (1991). Higher plant responses to environmental nitrate. *Physiologia Plantarum*, **82**, 640–50.

Redinbaugh, M.G. & Campbell, W.H. (1993). Glutamine synthetase and ferredoxin-dependent glutamate synthase expression in the maize (*Zea mays*) root primary response to nitrate. *Plant Physiology*, **101**, 1249–55.

Roche, D., Temple, S.J. & Sengupta-Gopalan, C. (1993). Two classes of differentially regulated glutamine synthetase genes are expressed in the soybean nodule: a nodule-specific class and a constitutively expressed class. *Plant Molecular Biology*, **22**, 971–83.

Sakakibara, H., Kawabata, S., Hase, T. & Sugiyama, T. (1992a). Differential effects of nitrate and light on the expression of glutamine synthetases and ferredoxin-dependent glutamate synthase in maize. *Plant Cell Physiology*, **33**, 1193–8.

Sakakibara, H., Kawabata, S., Takahashi, H., Hase, T. & Sugiyama, T. (1992b). Molecular cloning of the family of glutamine synthetase genes from maize: expression of genes for glutamine synthetase and ferredoxin-dependent glutamate synthase in photosynthetic and non-photosynthetic tissues. *Plant Cell Physiology*, **33**, 49–58.

Schuch, W. (1991). Using antisense RNA to study gene function. In *Molecular Biology of Plant Development*, ed. G.I. Jenkins & W. Schuch, pp. 117–27, Cambridge: Company of Biologists.

Shen, W.-J. (1991). Expression of glutamine synthetase genes from *Phaseolus vulgaris* L. in transgenic plants. PhD Thesis, Imperial College of Science and Technology, University of London.

Stock, J.B., Stock, A.M. & Mottonen, J.M. (1990). Signal transduction in bacteria. *Nature, London*, **344**, 395–400.

Sugiharto, B. & Sugiyama, T. (1992). Effects of nitrate and ammonium on gene expression of phosphoenolpyruvate carboxylase and nitrogen metabolism in maize leaf tissue during recovery from nitrogen stress. *Plant Physiology*, **98**, 1403–8.

Sugiharto, B., Suzuki, I., Burnell, J.N. & Sugiyama, T. (1992). Glutamine induces the N-dependent accumulation of mRNAs encoding phosphoenolpyruvate carboxylase and carbonic anhydrase in detached maize leaf tissue. *Plant Physiology*, **100**, 2066–70.

Temple, S.J., Knight, T., Unkefer, P.J. & Sengupta-Gopalan, C. (1993). Modulation of glutamine synthetase gene expression in tobacco by the introduction of an alfalfa glutamine synthetase gene in sense and antisense orientation: molecular and biochemical analysis. *Molecular and General Genetics*, **236**, 315–25.

Teverson, R. (1990). The differential expression of the genes encoding glutamine synthetase in developing root nodules. PhD thesis, University of Durham.

Tingey, S.V., Tsai, F-Y., Edwards, J.W., Walker, E.L. & Coruzzi, G.M. (1988). Chloroplast and cytosolic glutamine synthetase are encoded by homologous nuclear genes which are differentially expressed *in vivo*. *The Journal of Biological Chemistry*, **263**, 9651–7.

Tingey, S.V., Walker, E.L. & Coruzzi, G.M. (1987). Glutamine synthetase genes of pea encode distinct polypeptides which are differentially expressed in leaves, roots and nodules. *The EMBO Journal*, **6**, 1–9.

Vézina, L-P. & Langlois, J.R. (1989). Tissue and cellular distribution of glutamine synthetase in roots of pea (*Pisum sativum*) seedlings. *Plant Physiology*, **90**, 1129–33.

Vincentz, M., Moureaux, T., Leydecker, M-T., Vaucheret, H. & Caboche, M. (1993). Regulation of nitrate and nitrite reductase expression in *Nicotiana plumbaginifolia* leaves by nitrogen and carbon metabolites. *Plant Journal,* **3**, 315–24.

Wallsgrove, R.M., Keys, A.J., Bird, I.F., Cornelius, M.J., Lea, P.J. & Miflin, B.J. (1980). The location of glutamine synthetase in leaf cells and its role in the reassimilation of ammonia released in photorespiration. *Journal of Experimental Botany,* **31**, 1005–17.

Wallsgrove, R.M., Lea, P.J. & Miflin, B.J. (1979). Distribution of the enzymes of nitrogen assimilation within the pea leaf cell. *Plant Physiology,* **63**, 232–6.

Wallsgrove, R.M., Turner, J.C., Hall, N.P., Kendall, A.C. & Bright, S.W.J. (1987). Barley mutants lacking chloroplast glutamine synthetase – biochemical and genetic analysis. *Plant Physiology,* **83**, 155–8.

Walsh, K.B. (1990). Vascular transport and soybean nodule function. III. Implications of a continual phloem supply of carbon and water. *Plant, Cell and Environment,* **12**, 713–23.

PIERRE GADAL, LOÏC LEPINIEC,
and SIMONETTA SANTI

Interactions of nitrogen and carbon metabolism: implications of PEP carboxylase and isocitrate dehydrogenase*

Introduction

Over the last ten years, great progress has been made in understanding the biosynthesis of amino acids in plants. As far as incorporation of mineral nitrogen in organic compounds is concerned, the main interests have been focused on the enzymes implicated in nitrate and nitrite reduction (nitrate and nitrite reductases) and in ammonia assimilation (glutamine synthetase and glutamate synthase).

On the other hand, the net synthesis of amino acids requires the provision of carbon skeletons in the form of keto-acids. Two of the most important keto-acids used in amino acid biosynthesis, namely 2-oxoglutarate (2-OG) and oxaloacetate (OAA), are also TCA cycle intermediates. This paper considers the implications of phospho*enol*pyruvate (PEP) carboxylase (PEPC) and isocitrate dehydrogenase (ICDH) in the synthesis of these keto-acids.

PEP carboxylase

Phospho*enol*pyruvate carboxylase (PEPC) catalyses the β-carboxylation of phospho*enol*pyruvate in the presence of HCO_3^- to produce oxaloacetate and Pi (O'Leary, 1982). PEPC has been found in Eubacteria, green algae and higher plants (O'Leary, 1982). Many functions have been assigned to PEPC (Latzko & Kelly, 1983), the most extensively studied being its role in photosynthetic CO_2 fixation in C_4-plants. Nevertheless, the products that derive directly (oxaloacetate) or indirectly (aspartate, malate) from the action of PEPC are implicated in several cellular and metabolic events (tricarboxylic acid cycle replenishment, amino acid biosynthesis, recapture of respired CO_2, pH stat, ionic balance. . .).

The multifunctionality of enzymes is often correlated with multiple isoforms and the presence of PEPC isoenzymes is a general feature of

*This paper is dedicated to the memory of Claude CRÉTIN, without whom this paper and many others would never have been written.

most higher plants (Vidal & Gadal, 1983). More recently it has been demonstrated that these isoforms are encoded by small multigene families (Cushman *et al.*, 1989, Hudspeth & Grula, 1989, Crétin *et al.*, 1991, Hermans & Westhoff, 1992, Kawamura *et al.*, 1992). In Sorghum, two isoforms of PEPC, designated C_4- and C_3- types, have been characterized by their chromatographic and immunological properties (Vidal & Gadal, 1983).

PEPC gene family in Sorghum

A Sorghum λ EMBL4 genomic library was screened with a cDNA probe (pCP26B) (Thomas *et al.*, 1987) corresponding to the C_4- type isoenzyme. Three clones were distinguished by their restriction maps (Lepiniec *et al.*, 1993) as well as the intensity of hybridization; they were designated as CP 21, CP 28, and CP 46 and now, respectively, SvC3RI, SvC3 and SvC4. All of the three genes possess nine introns present at the same locations and encode a peptide chain of 960 amino acid residues. Specific probes from the 3' non-coding region of the three genes have been prepared and used to follow the specific expression of each gene in seedlings of Sorghum grown in the conditions described hereafter.

Influence of the nitrogen source on the expression of PEPC genes

To study the influence of the nitrogen source on PEPC activity and gene induction, *Sorghum vulgare* (Hybrid Tamaran) seeds were germinated for 3 days in the dark on a hydroponic solution containing 1 mM calcium sulphate in which oxygenation was maintained through air bubbling. Germinated plants were supplied with the nutrient solutions for macroelements described by Arnozis, Neleman and Findenegg (1988), transferred to light (350 µE M^{-2} s^{-1}), and harvested after the emergence of the second leaf (usually 4 days). The microelements were provided as specified in the Hoagland's solution.

Northern blot experiments using the gene-specific probes previously described, have shown that the three genes of PEPC carboxylase identified in Sorghum are differentially expressed in the different organs and tissues. In roots and leaves, CP28 does not seem to be regulated either by light or by the nitrogen source, consequently it is believed to be the housekeeping gene.

In leaves, CP46 is implicated in C_4 photosynthesis and its expression is induced by light (Thomas *et al.*, 1987). Furthermore, its expression is enhanced when the roots are fed with nitrate or ammonium ions, the effect of the latter seeming to be more pronounced. In all conditions tried so far, CP46 is not expressed in roots.

CP 21 is expressed in roots and leaves. In leaves, there is a slight induction by light but compared to CP46, this effect is negligible. In roots, the gene is clearly induced by ammonium ions but an effect by nitrate is barely detectable and could be due to a compound resulting from the reduction of the nitrate such as glutamine (Sugiharto *et al.*, 1992). In leaves, the influence of nitrogen source on CP21 expression is not obvious. Therefore, the expression of this gene is controlled by the nitrogen source mainly in roots.

Subsequent to the influence of the nitrogen source on the expression of the genes, the PEPC activity was determined in roots. The results are in good accordance with the data from Northern blots. The activity of the enzyme is increased both in the presence of ammonium ions and nitrate although in this case the effect is only around 20%.

Western blot experiments have confirmed the presence of higher amounts of PEPC protein in the ammonium-treated roots compared to the controls. The influence of the nitrogen source on the enzyme activities in leaves is presently being investigated.

Conclusion

In Sorghum, only one gene CP46 is organ-specific, in the sense that it is expressed only in leaves. The two other genes, CP 21 and CP 28, are expressed both in roots and leaves. CP28 is regulated neither by light nor by the nitrogen source; while CP21 is induced in roots by presence of ammonium ions. CP46 is induced in the leaves by light, the effect of which is enhanced by nitrate and ammonium ions.

Isocitrate dehydrogenase

2-Oxoglutarate represents the net source of carbon in the GS/GOGAT cycle. It has been recognized for a long time that two isocitrate dehydrogenases (ICDH) are present in higher plants: the NAD-ICDH (EC 1.1.1.41) restricted to the mitochondria, and the NADP-ICDH (EC 1.1.1.42) which has been found in the cytosol, mitochondria and chloroplasts (Chen & Gadal, 1990*a*, *b*).

Production of 2-oxoglutarate via the mitochondrial NAD-ICDH

It is widely accepted that this oxo-acid is derived from the Krebs cycle where it is produced through the action of the mitochondrial NAD-

ICDH (Bonner & Varner, 1965). The pathway is initiated by the oxidation of citrate and it is assumed that the resulting 2-oxo-acid traverses the mitochondrial and chloroplast membranes.

The evidence in favour of this scheme is: (a) the Krebs cycle is present in all higher plants (Douce, 1985) and is stimulated under conditions where there is a high demand for 2-oxoglutarate (Wegen et al., 1988); (b) translocators are present in the mitochondrial and chloroplast membranes (Douce, 1985; Woo, Flügge & Heldt, 1987), and therefore 2-oxoglutarate can gain access to the GOGAT enzyme located in the chloroplast stroma (Suzuki & Gadal, 1984).

Production of 2-oxoglutarate via the NADP-ICDH

Despite the evidence presented above, there are alternatives to the production of 2-oxoglutarate by mitochondrial NAD-ICDH. Higher plants contain another isocitrate dehydrogenase which also catalyses the oxidation of isocitrate to 2-oxoglutarate but which uses NADP rather than NAD. Recent studies have shown that there are at least three isoenzymes of NADP-ICDH: NADP-ICDH$_1$ localized in the cytosol, NADP-ICDH$_2$ localized in the chloroplast (Chen et al., 1989a) and NADP-ICDHm in the mitochondria (Rasmussen & Moller, 1990). From these studies it is apparent that the cytosol and the chloroplast represent two other sites for 2-oxoglutarate production.

Via the chloroplastic isoenzyme

Some years ago Elias and Givan (1977) and Randall and Givan (1981) suggested the presence of an NADP–ICDH activity associated with the chloroplast. Recently, we have confirmed this observation and shown that the NADP–ICDH$_2$ is different in its molecular and immunochemical properties from its cytosolic counterpart (Chen et al., 1989a). The purified enzyme exhibits a low K_m, for both NADP and isocitrate, in the mM range, and it is not affected by its product, 2-oxoglutarate, or amino acids such as glutamine and glutamate.

However, for the production of 2-oxoglutarate via this chloroplastic isozyme to be possible, two prerequisites must be met: the substrate must be available and the enzyme activity must be high enough to meet the demand for 2-oxoglutarate.

Concerning the first point, the source of isocitrate remains unknown. Aconitase activity has not been detected in this organelle and there is no evidence that chloroplasts produce this organic acid (Elias & Givan, 1977). As isocitrate is rather abundant in plant cells accounting, in some plants, for up to 85% of the total organic acids (Thimann &

Bonner, 1950) another possible source would be the importation of this metabolite from another compartment. However, to our knowledge, an efficient transport system for this tricarboxylic acid has yet to be demonstrated in the chloroplast envelope (Chen & Gadal, 1990*b*).

Green leaves have a high requirement for 2-oxoglutarate, because of the large amounts of chloroplastic glutamate exported to the cytosol (Woo *et al.*, 1987). In pea leaves, under our experimental conditions, the activity of glutamate synthase is about 57 mmol h^{-1} mg^{-1} Chl (Chen, & Gadal, 1990*b*). If the *in vitro* activity reflects the activity *in vivo* then the activity of the chloroplastic enzyme is not sufficient to produce the 2-oxoglutarate required for glutamate biosynthesis.

Via the cytosolic isoenzyme

The biogenesis of 2-oxoglutarate in the cytosol is supported by (a) the presence of a NADP-ICDH$_1$ in all plants examined (Chen *et al.*, 1989*b*); (b) the finding that NADP-ICDH$_1$ has a very low k_m value, in the μM range, for its substrates (Chen *et al.*, 1988); (c) the characterization of an active aconitase in the cytosol (Brouquisse *et al.*, 1987) – this point is of fundamental importance, since isocitrate in plant mitochondria is undetectable (Bowman, Ikuma & Stein, 1976); (d) the generation of citrate, the substrate for the aconitase, in the mitochondria (Barbareschi *et al.*, 1974), and its transportation to the cytosol via a tricarboxylate carrier (Day & Wiskich, 1984); (e) the finding of an alternative pool of citrate stored in the vacuole and moved to the cytosol by facilitated diffusion (Oleski, Mahdavi & Bennett, 1987); (f) the observation that an increase in the cytosolic NADP–ICDH activity coincides with the onset of nitrogen fixation in nodules of *Trigonella foenum-graecum* where a high supply of 2-oxoglutarate is required (Nautiyal & Modi, 1987).

The relative contribution of the three ICDHs

As discussed, the green plant cell contains three putative sites for 2-oxoglutarate synthesis: chloroplasts, cytosol and mitochondria. Due to its localization in the same compartment as GOGAT, the chloroplastic NADP–ICDH offers the advantage of producing 2-oxoglutarate at its site of utilization. However, the fact that the chloroplast NADP–ICDH$_2$ activity is low makes it unlikely to be significantly involved. As a consequence, the chloroplasts must rely on the import of 2-oxoglutarate from other pools. This hypothesis is supported by the existence of a very efficient carrier for this oxo-acid on the chloroplast envelope so that there is no limitation at this step (Proudlove, Thurman & Salisbury, 1984, Woo *et al.*, 1987).

For the production of 2-oxoglutarate used in glutamate biosynthesis, the mitochondrial NAD–ICDH displays severe disadvantages compared to the cytosolic NADP–ICDH: (a) in mitochondria the 2-oxoglutarate dehydrogenase is four times more active than the NAD–ICDH; it exhibits a low K_m for 2-oxoglutarate (in the 10 µM range) and therefore these properties strongly favour the channelling of the 2-oxoglutarate as a Krebs-cycle intermediate and consequently hinders its exportation out of the mitochondria; (b) the mitochondrial NAD–ICDH is 10 to 100 times less active (Ragland & Hackett, 1964) and exhibits an affinity for isocitrate which is 10-100 fold lower than that of the NADP–ICDH (Chen *et al.*, 1988); (c) the presence of a cytosolic aconitase makes isocitrate available for the NADP–ICDH$_1$ irrespective of the Krebs cycle, or alternatively without disrupting this fundamental pathway; (d) if produced inside the mitochondria, the oxo-acid must be transported through four membranes instead of two if made in the cytosol.

Conclusion

Three isocitrate dehydrogenases have been identified so far in higher plant cells, *i.e.* the mitochondrial NAD–ICDH, the chloroplastic and the cytosolic NADP–ICDHs. Of the three putative sites, the presently available evidences suggest that the cytosol constitutes the likely main site for 2-oxoglutarate production aimed at glutamate synthesis in the chloroplasts. Owing to the fundamental importance of the Krebs cycle in cell energetics, it appears logical that an alternative pathway operates outside the mitochondria for supplying large quantities of 2-oxoglutarate whenever it is needed.

Finally one must be cautious in the extrapolation *in vivo* of findings with isolated enzymes and further investigations using for instance, mutants or transgenic plants exhibiting reduced or increased levels of NADP–ICDH$_1$ and NADP–ICDH$_2$ are required to fully ascertain this point of view.

Acknowledgements

We wish to thank Jean VIDAL, Susana GALVEZ, Evelyne BISMUTH, Eliane KERYER, Ridong CHEN, Michaël HODGES and Valérie PACQUIT for their efficient contributions to experiments described in this chapter.

References

Arnozis, P.A., Neleman, J.A. & Findenegg, G.R. (1988). Phosphoen-olpyruvate carboxylase activity in plants grown with either nitrate or ammonium as inorganic nitrogen source. *Journal of Plant Physiology*, **132**, 23–7.

Barbareschi, D., Longo, G.P., Servettas, O., Zulian T. & Longo, C.P.(1974). Citrate synthetase in mitochondria and glyoxysomes of maize scutellum. *Plant Physiology*, **53**, 802–7.

Bonner, J. & Varner, J.E. (1965). The pathway of carbon in respiratory metabolism. In *Plant Biochemistry* ed. J. Bonner & J.E. Varner, pp. 211–12. New York: Academic Press.

Bowman, E.J., Ikuma, H. & Stein, H.J. (1976). Citric acid cycle activity in mitochondria isolated from mung bean hypocotyls. *Plant Physiology*, **58**, 426–32.

Brouquisse, R., Nishimura, M., Gaillard, J. & Douce, R. (1987). Characterization of a cytosolic aconitase in higher plant cells. *Plant Physiology*, **84**, 1402–30.

Chen, R.D., Bismuth, E., Champigny, M.L. & Gadal, P. (1989a). Chromatographic and immunological evidence that chloroplastic and cytosolic pea (*Pisum sativum* L.) NADP-isocitrate dehydrogenases are distinct isoenzymes. *Planta*, **178**, 157–63.

Chen, R.D., Bismuth, E., Issakidis, E., Pacot, C., Champigny, M.L. & Gadal, P. (1989b). Comparative immunochemistry of the NADP-isocitrate dehydrogenases: a highly conserved cytosolic enzyme in higher plants. *Comptes Rendus Académie des Sciences*, **308**, 459–65.

Chen, R.D & Gadal, P. (1990a). Structure function and regulation of NAD and NADP dependent isocitrate dehydrogenase in higher plants and in other organisms. *Plant Physiology and Biochemistry*, **28**, 411–27.

Chen, R.D. & Gadal, P. (1990b). Do the mitochondria provide the 2-oxoglutarate needed for glutamate synthesis in higher plants chloroplasts? *Plant Physiology and Biochemistry*, **28**, 141–5.

Chen, R.D., Le Maréchal, P., Vidal, J., Jacquot, J.P. & Gadal, P. (1988). Purification and comparative properties of the cytosolic isocitrate dehydrogenases (NADP) from pea (*Pisum sativum*) roots and green leaves. *European Journal of Biochemistry*, **175**, 565–72.

Crétin, C., Santi, S., Keryer, E., Lepiniec, L., Tagu, D., Vidal, J. & Gadal, P. (1991). The phosphoenolpyruvate carboxylase gene family in Sorghum: promoter structure, amino acid sequences and expression of genes. *Gene*, **99**, 87–94.

Cushman, J.C., Meyer, G., Michalowski, C.B., Schmitt, J.M. & Bohnert, H.J. (1989). Salt stress leads to differential expression of two isogenes of phosphoenolpyruvate carboxylase during crassulacean acid metabolism induction in the common ice plant. *The Plant Cell*, **1**, 715–25.

Day, D.A. & Wiskich, J.T. (1984). Transport processes of isolated plant mitochondria. *Physiologie Végétale*, **22**, 241–61.

Douce, R. (1985). *Mitochondria in Higher Plants. Structure, Function, Biogenesis.* Orlando: Academic Press.

Elias, B.A. & Givan, C.V. (1977). Alpha-ketoglutarate supply for amino acid synthesis in higher plant chloroplasts. *Plant Physiology*, **59**, 738–40.

Hermans, J. & Westhoff, P. (1992). Homologous genes for the C4 isoform of phosphoenolpyruvate carboxylase in a C3 and a C4 *Flaveria* species. *Molecular and General Genetics*, **234**, 275–84.

Hudspeth, R.L. & Grula, J.W. (1989). Structure and expression of the maize gene encoding the phosphoenolpyruvate carboxylase isozyme involved in C4 photosynthesis. *Plant Molecular Biology*, **12**, 579–89.

Kawamura, T., Shigesada, K., Toh, H., Okumura, S., Yanagisawa, S. & Izui, K. (1992). Molecular evolution of phosphoenolpyruvate carboxylase for C4 photosynthesis in maize: comparison of its cDNA sequence with a newly isolated cDNA encoding an isozyme involved in the anaplerotic function. *Journal of Biochemistry*, **112**, 147–54.

Latzko, E. & Kelly, G. (1983). The many-faceted function of phospho-enolpyruvate carboxylase in C3 plant. *Physiologie Végétale*, **215**, 805–15.

Lepiniec, L., Kéryer, E., Philippe, H., Gadal, P. & Crétin, C. (1993). Sorghum phosphoenolpyruvate carboxylase gene family: structure, function and molecular evolution. *Plant Molecular Biology*, **21**, 487–502.

Nautiyal, C.S. & Modi, V.V. (1987). Malate dehydrogenase and isocitrate dehydrogenase in root nodules of *Trigonella*. *Phytochemistry*, **26**, 1863–5.

O'Leary, M.H. (1982). Phosphoenolpyruvate carboxylase: an enzymologist's view. *Annual Review of Plant Physiology*, **32**, 297–315.

Oleski, N., Mahdavi, P. & Bennett, A.B. (1987). Transport properties of the tomato fruit tonoplast. *Plant Physiology*, **84**, 997–1000.

Proudlove, M.O., Thurman, D.A. & Salisbury, J. (1984). Kinetic studies on the transport of 2-oxoglutarate and L-malate into isolated pea chloroplasts. *New Phytologist*, **96**, 1–5.

Ragland, T.E. & Hackett D.P., (1964). Compartmentation of nicotinamide dinucleotide dehydrogenases and transhydrogenases in non-photosynthetic plant tissues. *Archives of Biochemistry and Biophysics*, **108**, 479–89.

Randall, D.D. & Givan, C.V. (1981). Subcellular location of NADP-isocitrate dehydrogenase in *Pisum sativum* leaves. *Plant Physiology*, **68**, 70–3.

Rasmussen, A.G. & Moller, I.M. (1990). NADP-utilizing enzymes in the matrix of plant mitochondria. *Plant Physiology*, **94**, 1012–18.

Sugiharto, B., Susuki, I., Burnell, J.N. & Sugiyama, T. (1992). Gluta-mine induces the N-dependent accumulation of mRNAs encoding phosphoenolpyruvate carboxylase and carbonic anhydrase in detached maize leaf tissue. *Plant Physiology,* **100,** 2066–70.

Suzuki, A. & Gadal, P. (1984). Glutamate synthase: physicochemical and functional properties of different forms in higher plants and in other organisms. *Physiologie Végétale,* 22, 471–86.

Thimann, K.V. & Bonner, W.D. Jr. (1950). Organic acid metabolism. *Annual Review of Plant Physiology,* 1, 75–108.

Thomas, M., Crétin, C., Kéryer, E., Vidal, J. & Gadal, P. (1987). Photocontrol of Sorghum leaf phosphoenolpyruvate carboxylase: characterization of messenger RNA and of photoreceptor. *Plant Physiology,* **85,** 243–6.

Vidal, J. & Gadal, P. (1983). Influence of light on phosphoenolpyruvate carboxylase in Sorghum leaves. Identification and properties of two isoforms. *Physiologia Plantarum,* **57,** 119–23.

Wegen, H.G., Birth, D.G., Elrifi, I. & Turpin, D.H. (1988). Ammonium assimilation requires mitochondrial respiration in the light. *Plant Physiology,* **86,** 688–92.

Woo, K.C., Flügge, U.I. & Heldt, H.W. (1987). A two-translocator model for the transport of 2-oxoglutarate and glutamate in chloro-plasts during ammonia assimilation in the light. *Plant Physiology,* **84,** 624–32.

M. JACOBS, V. FRANKARD, M. GHISLAIN
and M. VAUTERIN

The genetics of aspartate derived amino acids in higher plants

Introduction

Although the central role of amino acid biosynthesis in plant metabolism and development is evident, progress in the understanding of the molecular mechanisms by which such pathways are regulated is only recent. To elucidate the genetic regulation of a pathway, mutants have proved to be invaluable in the analysis of microbial systems. The same is true for plants, but again there are only a few pathways for which mutants with defective or altered essential functions are available. However, the improvements in plant mutant isolation at cell and whole plant levels, and in molecular biology have now provided mutants affected in amino acid metabolism as well as amino acid biosynthetic genes or cDNAs (for a review, see Last, 1992).

The branched biosynthetic pathway of aspartate-derived amino acids (lysine, methionine, threonine) and the related pathway leading to isoleucine, valine and leucine have received much interest for various reasons. First, it gives rise to essential amino acids which, if poorly represented as lysine in cereals and methionine in legumes, limit the nutritional quality of crop plants as diet for human beings and monogastric animals. Secondly, three classes of potent herbicides (sulphonylureas, imidazolinones and triazolopyrimidin) kill plants through the inhibition of acetolactate synthase, an enzyme common to the isoleucine and valine pathways. Key regulatory enzymes of the aspartate-derived amino acid biosynthesis would also be suitable targets for efficient herbicides while little affecting the environment. Finally, this pathway, subject to complex and coordinated biochemical control, would provide new insights into the mechanisms involved in the regulation of amino acid biosynthesis. As a result, we are dealing with a pathway for which there are now a whole array of mutants, a series of enzymes characterized at the biochemical, and for some of them at the molecular, level and a few genes or c-DNAs allowing transgenic analysis.

In this review, we will emphasize genetic and molecular studies of the aspartate-derived amino acid pathway, and try to indicate what type of genetic manipulations of this metabolic network can lead to carbon flux towards the lysine biosynthetic branch no longer limited by the feedback–inhibition type of regulation.

General features of the aspartate pathway

The aspartate family, comprising aspartate, asparagine, lysine, threonine, isoleucine, and methionine, has as a major carbon donor, oxaloacetate, from the citric acid cycle, which after transamination yields aspartate. Pyruvate also contributes to the lysine and isoleucine biosynthesis. Fig. 1 gives the main features of the aspartate family amino acid biosynthesis pathway in bacteria and plants. Whilst in bacteria both

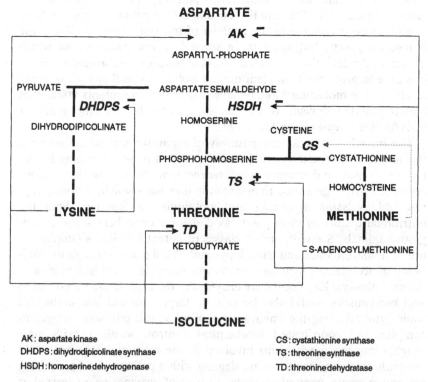

AK: aspartate kinase
DHDPS: dihydrodipicolinate synthase
HSDH: homoserine dehydrogenase

CS: cystathionine synthase
TS: threonine synthase
TD: threonine dehydratase

Fig. 1. The aspartate-derived amino acid biosynthetic pathway and its main regulatory steps. Feedback inhibition: (−); feedback activation: (+); repression/induction: − − − − −.

induction and repression of enzyme synthesis by pathway products and feedback inhibition play a role in the regulation (Datta, 1969), the major method of regulation of the pathway in plants appears to be feedback inhibition at the level of the key enzymes, aspartate kinase (AK), the first enzyme of the pathway, dihydrodipicolinate synthase (DHDPS), the first enzyme of the lysine branch, and homoserine dehydrogenase (HSDH), the first enzyme of the branch leading to threonine and methionine.

Generally, AK is inhibited by lysine, by threonine and also synergistically by lysine plus *S*-adenosylmethionine, while DHDPS is the only enzyme of this branch which is strongly feedback inhibited by lysine. Threonine-sensitive and threonine-insensitive isoforms of HSDH have been identified in plants. Stimulation of the threonine synthase activity in presence of *S*-adenosylmethionine has been reported for all plants studied to date. Cystathionine synthase is the only enzyme for which variations in the level of enzyme activity by induction vs. repression in response to the methionine pool in the cell have been published (Rognes *et al.*, 1986; Thompson *et al.*, 1982). Table 1 summarizes our present knowledge concerning the regulation of key enzymes of the aspartate-derived amino acid pathway.

Many enzymes of the pathway encoded by nuclear genes have been localized in the chloroplast. Only methionine synthase, which converts homocysteine to methionine, appears to be localized in the cytoplasm (Wallsgrove, Lea & Miflin, 1983). The amino acid pool in the chloroplast thus determines the feedback regulatory properties of the pathway.

Mutants of aspartate kinase and dihydrodipicolinate synthase: genetic and biochemical analysis

We have seen that the main limitation of the entry of the carbon flux into the different branches of the aspartate-derived amino acid biosynthetic pathway occurs at the level of key enzymes, aspartate kinase (AK) for threonine synthesis and dihydrodipicolinate synthase (DHDPS) for lysine synthesis. With the goal of enhancing threonine and lysine in the free amino acid pool, such enzymes have to be modified in a way that their catalytic activity is increased and/or that their sensitivity to the feedback inhibition by the end-products is decreased.

Based on experience derived from mutant isolation in prokaryotes and lower eukaryotes, attempts to isolate mutants in higher plants using amino acid analogues or growth inhibitory concentrations of amino acids were carried out starting in the late seventies. Whole

Table 1. *Regulatory steps in the aspartate pathway in higher plants*

Enzyme	Isozymes	Effector(s)	Gene or cDNA	Branch
By feedback regulation				
Aspartate kinase	≥ 2	Lysine (−) Threonine (−) Lysine plus AdoMet (−)	*ak–hsdh* (carrot, maize, *A. thaliana*)	Common
Homoserine dehydrogenase	≥ 1	Threonine (−)	Idem	Threonine, Methionine
Dihydrodipicolinate synthase	1	Lysine (−)	*dhps* (maize, wheat, poplar, soybean, *A. thaliana*, *N. sylvestris*)	Lysine
Threonine synthase	1	AdoMet (+)	NA	Threonine
Threonine deaminase	1	Isoleucine (−)	*td* (tomato)	Isoleucine
Acetolactate synthase	≥ 1	Leucine(−) Valine (−)	*als* (*A. thaliana, N. tabacum, B. napus*)	Isoleucine
By induction/ repression				
Cystathionine synthase	1	Methionine (+,−)	NA	Methionine

NA: not available.
−: feedback inhibition or repression. +: feedback stimulation or induction.

seeds, excised embryos, embryoids, calli and protoplasts were muta-
genized, and thereafter submitted to a particular selection (and regeneration if possible). If an increase in the specific activity of a particular
enzyme was not unambiguously reported for the aspartate pathway,
the modification by mutation of feedback inhibition properties was on
the contrary successful.

Growth inhibition when lysine plus threonine (Lys + Thr) were
present in the culture medium, which could be restored if methionine

or its precursors were added, led to the assumption that aspartate kinase isozymes were strongly feedback-inhibited by these two amino acids, thus preventing the formation of sufficient aspartylphosphate for methionine synthesis. Mutants resistant to Lys + Thr have been isolated in maize (Hibberd *et al.*, 1980; Hibberd & Green, 1982; Miao, Duncan & Widholm, 1988; Diedrick, Frisch & Gengenbach, 1990; Dotson *et al.*, 1990), barley (Cattoir-Reynaerts, Degryse & Jacobs, 1981; Bright *et al.*, 1982*a*; Bright, Miflin & Rognes, 1982*b*), carrot (Cattoir-Reynaerts *et al.*, 1983), *Arabidopsis thaliana* (Vernaillen *et al.*, 1985) and *Nicotiana sylvestris* (Negrutiu *et al.*, 1984). In almost all cases, inheritance of the resistance trait was monogenic, dominant and nuclear. Furthermore, in barley (Bright *et al.*, 1982*a*) and in maize (Diedrick *et al.*, 1990), distinct mutants presenting unlinked loci for the resistance property were isolated. At the enzymatic level, a reduced sensitivity of aspartate kinase activity to feedback inhibition by lysine was consistently observed in all of these mutants. In particular, mutant RLT70 in *N.sylvestris* contained an isozyme totally desensitized to feedback inhibition by lysine, as compared to the wild type in which 80% of the total AK activity is inhibited by this effector (Fig. 2). As a consequence, the free amino acid composition of all tissues analysed, including seeds, was modified: free threonine content was markedly increased (in leaves, up to 70% of the total free pool, against 9% in the wild type), free isoleucine also accumulated significantly, as well as lysine and methionine but to a lesser extent (Table 2). Impact on

a) RAEC-1 mutant: DHDPS activity   **b) RLT 70 mutant: AK activity**

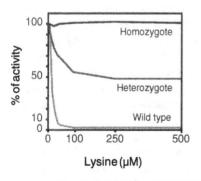
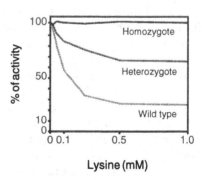

Fig. 2. Inhibition of DHDPS and AK activity by lysine in *N.sylvestris* heterozygote and homozygote mutants RAEC-1 and RLT-70.

Table 2. *Comparison of the free content of amino acids derived from aspartate in leaves from mature* Nicotiana sylvestris *wild type plants, AEC-resistant mutant, Lys+Thr resistant mutant and hybrid resulting from the cross of the two preceding mutants (expressed in absolute values in nmoles/g fresh weight and in relative values as percentage of the total)* (Frankard et al., 1992)

	Wild type		AEC-resistant		LT-resistant		Hybrid	
Amino acid	nmoles/g	%	nmoles/g	%	nmoles/g	%	nmoles/g	%
Aspartate	694	14.6	1042	16.3	1977	6.6	1286	21.7
Threonine	335	7.1	159	2.5	18 779	62.9	51	0.9
Methionine	22	0.5	30	0.5	184	0.6	5	0.1
Isoleucine	83	1.8	78	1.2	752	2.5	17	0.7
Lysine	92	1.9	1380	21.6	697	2.3	1785	30.1
Total	4741		6395		29 872		5932	

total (free + bound) threonine content in leaves was six-fold and in seeds two-fold. The regenerated plants displayed in most of the cases a normal phenotype and were fertile (Frankard *et al.*, 1991).

Among several lysine analogues studied for their ability to block plant growth, *S*-(2-aminoethyl)L-cysteine (AEC) was the most efficient. AEC appears to be incorporated into proteins instead of lysine, implying that lysine aminoacyl-tRNA poorly discriminates between lysine and its analogue. AEC also can inhibit enzymes that are normally feedback inhibited by lysine, such as AK and DHDPS, although a 10-fold higher concentration is needed to reach the same inhibition. A dilution-type mechanism of resistance should counter the effect of the analogue. Using this selection agent, several mutants have been obtained in which the resistance could be ascribed to different causes: decreased uptake, in one case lysine overproduction, and other mechanisms unidentified to date.

In barley (Bright, Featherstone & Miflin, 1979*a*; Bright, Norbury & Miflin, 1979*b*), carrot (Matthews, Shye & Widholm, 1980), *Arabidopsis thaliana* (Cattoir-Reynaerts *et al.*, 1981), wheat (Kumpaisal, Hashimoto & Yamada, 1988) and rice (Miao *et al.*, 1988), recessive mutants resistant to AEC were shown to have reduced root uptake of both lysine and AEC. One possible explanation for the repeated isolation of uptake mutants from this AEC selection procedure may be the presence in these species of a limiting feedback control at the level of AK, so that a dilution-type mechanism of resistance would require a double mutation at the level of both key enzymes, DHDPS and AK.

In rice (Shaeffer & Sharpe, 1981,1987, 1990), in maize (Gengenbach, 1984) and in potato (Jacobsen, 1986), resistant cell lines have been isolated but with no modification of the free amino acid pool, of the lysine content in particular, and without impairing the lysine uptake system. The cause for AEC resistance is unclear. One monocot, *Pennisetum americanum*, selected as AEC-tolerant (Boyes & Vasil, 1987) did accumulate free lysine but no genetic or biochemical data were available. In *Nicotiana sylvestris*, Negrutiu *et al.* (1984) regenerated from protoplasts grown in the presence of AEC, plants that produced up to 30% free lysine in leaves, against 2% in the wild type (Table 2). The resistance to AEC in the mutant RAEC-1 was inherited as a monogenic, dominant and nuclear trait. Enzymatic analysis revealed clearly that DHDPS was totally desensitized to the lysine feedback inhibition in the homozygous plant (Fig. 2). However, mutants characterized by high levels of free lysine (more than 20% of the total free amino acid content) were associated with an aberrant phenotype, resulting in reduced leaf blade size, absence of apical dominance, and sterility. Free lysine accumulation in seeds was not observed.

In *Nicotiana sylvestris*, a double mutant was obtained by crossing a Lys+Thr resistant plant with the mutant RAEC-1 (Frankard, Ghislain & Jacobs, 1992). In other words, both AK and DHDPS partially desensitized to lysine feedback inhibition were simultaneously present in this hybrid plant. The most remarkable feature of this hybrid heterozygote for the mutated AK and DHDPS genes was a very high content of free lysine (up to 50% of the total free amino acid content against up to 30% in the homozygote AEC resistant plant and 2% in the wild type plant, see Table 2), at the expense of the other aspartate-derived amino acids, but always accompanied with the extremely aberrant phenotype described for some of the RAEC-1 mutants (Fig. 3). This suggests that lysine, in contrast to threonine, is involved in growth and differentiation processes yet to be demonstrated. As a conclusion, though threonine overproduction does not seem to greatly affect plant development, the effect of lysine overproduction on plant morphology points out the need to control free lysine levels which could be reached in future transgenic plants.

(a)

Fig. 3(a). Aberrant phenotype presented by the *N.sylvestris* hybrid RAEC-1 × RLT-70, when high free lysine content is encountered in leaves, compared to the wild-type (b).

Cloning and characterisation of plant genes coding for enzymes of the aspartate pathway

The genes of interest in plants are those coding for the carbon flux-limiting enzymes AK and DHDPS and their mutated alleles coding for the forms desensitized to lysine feedback inhibition. Molecular cloning of AK has been achieved first in carrot (Weisemann & Matthews, 1993) and then in soybean (Gebhardt, Weisemann & Matthews, 1993), *Arabidopsis thaliana* (Ghislain *et al.*, 1994) and maize (Muehlbauer *et*

(*b*)

Fig. 3(b).

al., 1993). The deduced amino acid sequences from these genes have a marked identity with the *E.coli* proteins aspartate-kinase I-homoserine dehydrogenase I and aspartate-kinase II-homoserine dehydrogenase II, which are bifunctional enzymes performing the first and third step of the aspartate pathway (Table 3). It has been proposed that the protein is divided into three domains: an amino-terminal domain with AK

Table 3. *Sequence similarities of the AK-HSDH bifunctional enzymes (expressed as percentage of the apoprotein amino acid sequence)*

	E.c. AKI–HSDHI	E.c. AKII–HSDHII	D.c. AK–HSDH	A.t. AK–HSDH
E. coli AKI–HSDHI	100			
E. coli AKII–HSDHII	29.8	100		
D. carota AK–HSDH	37.2	31.4	100	
A. thaliana AK–HSDH	37.5	30.7	80.3	100

activity (around 250 amino acids), an intermediate domain with no known catalytic activity but most likely involved in the regulation (around 220 amino acids), and finally a carboxy-terminal domain with HSDH activity (around 350 amino acids). A closer look at the homology between the plant genes cloned and the two *E.coli* bifunctional proteins reveals a higher degree of identity with AK I-HSDH I, which is feedback-controlled by threonine (and thus naturally insensitive to lysine). Southern blot analysis with genomic DNA from carrot and *A.thaliana* revealed a simple banding pattern (Fig. 4), whereas multiple bands were observed for soybean, and at least four different clones were distinguished in maize based on their restriction analysis. In all *ak–hsdh* plant cDNAs characterized to date, a transit peptide rich in serine, threonine and small hydrophobic amino acids has been found, in accordance with the earlier localization of the AK and HSDH activities in the chloroplast (Wallsgrove, Lea & Miflin, 1983). The beginning of the apoprotein has not yet been identified, but must occur before the highly conserved amino acid stretch KFGG, found in all *ak* genes cloned to date.

In *A.thaliana,* characterization of a genomic clone led to the identification of a total of 17 introns ranging between 78 and 134 bp: 15 in the apoprotein coding region, one in the transit peptide, and one in the 5′ untranslated leader sequence of the gene (Fig. 5). The whole gene product is made of 892 residues and has a calculated molecular weight of 90 kDa. Two putative functional amino acid conserved sequences have been found in the mature protein: the D–P–R sequence which is probably involved in the kinase activity, and the G–X–G–X–

Fig. 4. Southern blot analysis of *A.thaliana*, using the carrot *ak–hsdh* gene as probe (lane 1: *Bam*HI; lane 2: *Eco*RI; lane 3: *Hin*dIII; lane 4: control).

X–G sequence which is most probably the NADPH binding domain of homoserine dehydrogenase. The translational start has been localized after isolation of a full length cDNA. Using a 270 bp fragment of the 5′ upstream region comprising putative TATA and CAAT boxes fused to GUS as reporter gene in transient expression experiments, promoter-like activity of this sequence could be demonstrated (Figs. 6 and 7). Interestingly, two possible regulatory elements were also identified in the 5′ upstream region: an *opaque-2*-binding site recognized in maize by a transcriptional factor regulating storage protein synthesis, and a GCN4 sequence involved in yeast in the general control of amino acid biosynthesis (Fig. 6). Further investigation should reveal if these novel types of regulation for amino acid biosynthesis in higher plants are actually functional. The 3′ flanking sequence presented two putative polyadenylation signals 25 bp apart, shortly followed by the polyadenylation site (Ghislain *et al.*, 1994).

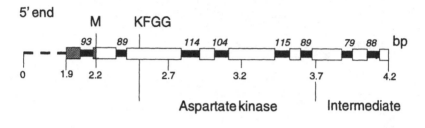

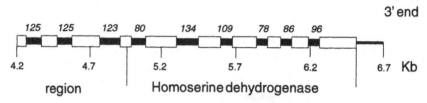

Fig. 5. Genomic structure of the *ak–hsdh* gene of *A.thaliana* (white boxes: exons; grey boxes: 5′ untranslated region; black boxes: introns. M: 1st ATG; KFGG: highly conserved amino acid motif).

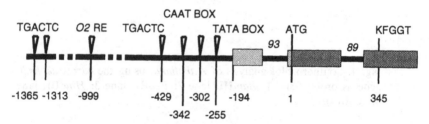

Fig. 6. 5′ Upstream region of the *ak–hsdh* gene of *A.thaliana*, and the different putative promoter and regulatory elements. O2RE: *opaque2* recognition element; TGACTC: GCN4 binding site.

It is important to underline the fact that, to date, the gene coding for the lysine sensitive isozyme of AK (for which mutated alleles exist) has not yet been cloned. Southern blot analysis on genomic DNA from *A.thaliana* probed with degenerated oligonucleotides corresponding to the highly conserved N-terminal amino acid sequence KFGG have revealed extra bands in addition to the ones corresponding to the plant *thrA* equivalent. Northern blot analyses are underway to check for the presence of corresponding messenger RNAs.

The gene which encodes DHDPS has firstly been cloned in monocots, wheat (two isozymes; Kaneko *et al.*, 1990) and maize (Frisch

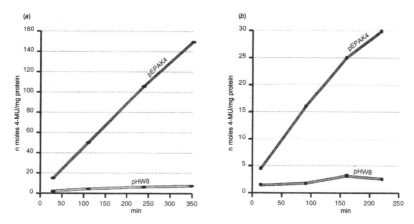

Fig. 7. Transient expression of GUS fused to a 270bp sequence of the 5' upstream region of the *ak–hsdh* gene of *A.thaliana* (comprising the putative promoter elements, pEPAK4), in protoplasts from mesophyll (*a*) and suspension culture (*b*) cells. pHW8 is a promoterless construct also containing GUS.

et al., 1991), and later on in soybean (Silk & Matthews, 1993) and *N.sylvestris* (Ghislain, personal communication). There appears to be a unique DHDPS per genome (Fig. 8). Sequence analysis reveals a certain divergency between the DHDPS of monocots and dicots (Fig. 9), but a long internal stretch of amino acids is conserved among the different plant *dhdps* genes (Table 4). At the 5' end, a transit peptide, with the characteristics of plastid sequences, has been identified in all plants, which was expected from the subcellular location of the enzyme activity. Promoters of the *dhdps* genes from soybean and *A.thaliana* are now being characterized (Silk & Vauterin, pers.comm.). Molecular weights deduced from the different sequences range from 35 700 to 35 900, which is in accordance with molecular weights of DHDPS determined biochemically (homotetramer of ± 38 000). Concerning the 3' region, in wheat as well as in *N.sylvestris,* only derivatives of the proposed polyadenylation signal (AATAAA) were present downstream of the stop codon.

In *N.sylvestris* RAEC-l mutant, the gene coding for the lysine-insensitive DHDPS was cloned and sequenced in order to identify the mutation conferring this trait (Ghislain, pers.comm.). Two nucleotides were actually substituted in the mutated allele, changing an asparagine to an isoleucine in a highly conserved region (Table 5). Interestingly,

(a)　　　　　　(b)　　　　　　(c)

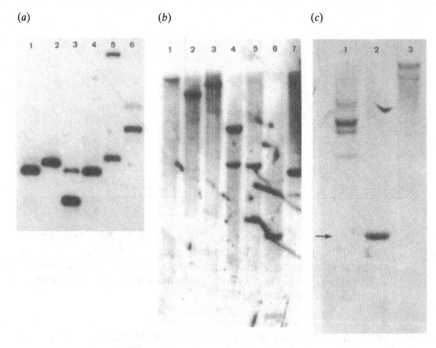

Fig. 8. Southern blot analysis of poplar (a) (1: *Bam*HI; 2: *Eco*RI; 3: *Hind*III; 4: *Xba*I; 5: *Pst*I; 6: *Hind*II), *A.thaliana* (b) (1: *Eco*RI; 2: *Bam*HI; 3: *Pst*I; 4: *Acc*I; 5: *Nco*I; 6: *Hinc*II; 7: *Sac*I), and *N.sylvestris* (c) (1: *Hind*III; 2: *Eco*RI; 3: *Bam*HI) using respective *dhdps* cDNA or genes as probes.

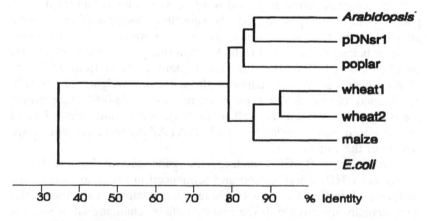

Fig. 9. Dendogram obtained by UPGMA clustering of the different *dhdps* nucleotide sequences.

Table 4. *Amino acid sequence identity between the different plant* dhdps *genes available (expressed as percentage of the amino acid sequence of the mature protein)*

Arabidopsis thaliana	100					
Nicotiana sylvestris	80.2	100				
Poplar	70.9	79.2	100			
Wheat 1	86.8	73.7	65.0	100		
Wheat 2	86.2	71.3	64.3	81.2	100	
Maize	67.4	70.9	68.7	76.1	75.3	100
	A.t.	*N.s.*	Poplar	Wheat 1	Wheat 2	Maize

Table 5. *Localized nucleotide substitutions in mutated DHDPS less sensitive to lysine feedback inhibition*

Plant	Mature peptide amino acid residue	Nucleotide change	Amino acid change	Reference
Maize	103	G to A	Ser to Asn	Gengenbach (p.c.)
	108	G to A	Glu to Lys	id
	112	C to T	Ala to Val	id
	112	G to A	Ala to Thr	id
Nicotiana sylvestris	104	AC to TT	Asn to Ile	Ghislain (p.c.)

desensitized maize DHDPS obtained after *in vitro* random mutagenesis also presented nucleotide substitutions in the same conserved region (Gengenbach, pers.comm.). After functional complementation of an *E.coli* strain lacking DHDPS activity with the mutated plant gene, an enzymatic test confirmed the insensitivity of the enzyme to feedback inhibition by lysine (Table 6). Study of this gene's expression throughout plant growth and in the different organs is presently underway.

Table 6. *Lysine inhibition of DHDPS activity in plants and in an* E.coli dapA- *strain functionally complemented with the* dhdps-rl *mutated allele (in arbitrary units)*

L-lysine	DHDPS, wild type N.sylvestris	DHDPS, RAEC-1 N.sylvestris	E.coli dapA-, with pDNsr1
0 μM	660	492	644
100 μM	34	470	695

Genetic and molecular manipulation of the regulatory network of the aspartate pathway

The availability of the genes coding for key enzymes of the pathway allows their manipulation in transgenic plants harbouring chimaeric constructs. For example, mutated c DNAs leading to overproduction of lysine or/and threonine can be placed under the control of an array of constitutive or tissue-specific regulatory elements. However, the genes described in Part III were isolated only very recently and obtaining transformants expressing the cloned plant coding sequences and regulatory elements, although being a concrete objective of research, is still in a preliminary phase.

As an alternative, corresponding bacterial genes were introduced into tobacco and potato under the control of plant-recognized regulatory sequences and with adequate polyadenylation signal sequences and the resultant effects on amino acid biosynthesis determined. Table 7 shows the wild type or mutated genes which have been used in such an approach.

Glassman (1992) associated the *dapA* gene of *E.coli* (which codes for a DHDPS at least 200-fold less sensitive to inhibition by lysine *in vitro* compared to the plant enzyme), the transit peptide of the small subunit of ribulose biphosphate carboxylase of pea and the 3' polyadenylation signal sequence from the nopaline synthase gene. This cassette was placed under the control of the 35S promoter of CaMV and inserted via *Agrobacterium* transformation in *Nicotiana tabacum*. Putative transgenic plants were screened by evaluating the response of leaf disks plated on a medium supplemented with AEC. AEC-tolerant plants displayed high levels of free lysine in leaves associated with an aberrant morphology. High levels of lysine-tolerant DHDPS activity were observed in extracts of leaves and seeds. However, although the free lysine pool is significantly increased in transgenic plants expressing

Table 7. *Transfer and expression of heterologous genes*

Bacterial genes		
E.coli	*dap*A	DHDPS 20 to 100 fold less sensitive to lysine inhibition than plant enzymes[a,b]
	*lys*C TOC R21	lysine insensitive mutated AKIII[c]
	*thr*A[r]	threonine insensitive mutated AKI-HSDHI[d]
Plant genes		
N.sylvestris	*dhdps-r1*	DHDPS insensitive to lysine inhibition
A.thaliana	*ak–hsdh*	threonine sensitive, lysine insensitive AK–HSDH

[a]Glassman *et al.*, 1992; [b]Shaul and Galili, 1992; [c]Shaul and Galili, 1992; [d]Colau, 1988;

a lysine-tolerant DHDPS, no elevation in lysine level could be detected in seeds.

The Laboratory of Plant Genetics at the Weizmann Institute of Science, has recently produced transgenic tobacco and potato plants expressing bacterial DHDPS and AK enzymes, which are both less sensitive to inhibition by lysine than the plant enzymes (Shaul & Galili, 1992*a*, *b*; Perl, Shaul & Galili, 1992). Chimeric genes included the 35S promoter and Ω DNA sequence coding for the Ω mRNA leader sequence with or without the transit peptide of the pea ribulose biphosphate carboxylase gene. Leaves of both types of transgenic plants exhibited a significant overproduction of free lysine or free threonine, respectively (Fig. 10). Accumulation of the essential amino acid was, in each case, positively correlated with the level of activity of the corresponding enzyme in a series of transgenic plants.

Crop seeds generally contain very low levels of free threonine and lysine, amino acids which are often present in low amounts in seed protein. In the previously mentioned transformants, if a significant elevation of threonine content (up to 14-fold) was observed in seeds, no accumulation of lysine could be detected, as reported by Glassman (1992). Another major drawback of these transformants was the relationship between overproduction and abnormal morphology, especially in the case of lysine.

Therefore, the authors used chimeric constructs containing the same bacterial genes but now fused to a promoter from a phaseolin seed storage protein gene. The expression of the mutated *lysC* coding for

(*a*)

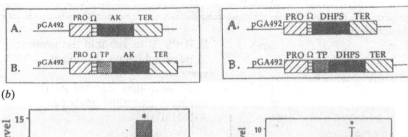

(*b*)

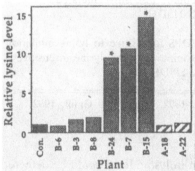 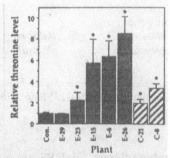

Fig. 10. (*a*) Schematic diagram of the chimeric gene including DHDPS (the coding sequence of the *E.coli dap*A gene) or AK (the coding sequence of the mutant *E.coli lys*C allele). A: Cytoplasmic-type gene. B: Chloroplastic-type gene. PRO: The CaMV 35S promoter. Ω: The DNA coding for the Ω mRNA leader sequence. TER: The DNA sequence of the octopine synthase 3′ termination. (*b*) Relative levels of free lysine or free threonine in leaves of transgenic tobacco plants, plotted relative to the level of each free amino acid in the control plants. Con: control as non-transformed plants. B-3 to B-24, E-6 to E-24: transgenic plants expressing the chloroplastic-type of DHDPS and AK, respectively. A-18 and A-22, C-8 and C-21: transgenic plants expressing the cytoplasmic-type of DHDPS and AK, respectively (according to Shaul & Galili, 1992*a, b*)

a desensitized AK was accompanied by a significant increase of threonine and methionine in seeds, without alteration of the phenotype (Karchi, Saul & Galili, 1993). However, in the case of the *dapA* transformants, the accumulation of free lysine increased only slightly at early stages of seed development and declined to the level of control plants at maturity (Karchi, Shaul & Galili, 1994). The same authors proposed that the failure of accumulation in seeds was due to enhanced catabolism by lysine ketoglutarate reductase, an enzyme metabolizing lysine into saccharopine. In fact, the activity of lysine ketoglutarate reductase was significantly higher in the seeds of the transgenic plants expressing DHDPS than in the controls. As already mentioned for the hybrid between RAEC-1 and RLT-70, they also observed that, in transgenic plants co-expressing the two bacterial enzymes, lysine synthesis was increased further at the expense of threonine, but without resulting in lysine accumulation into mature seeds.

Conclusions and perspectives

Thus, to date, results from both the analysis of lysine and threonine overproducing mutants and transgenic plants transformed with bacterial genes coding for AK and DHDPS clearly indicate that a constitutive expression of the foreign genes leads to higher threonine content in leaves or seeds, but for lysine the increase appears only in leaves correlated to an abnormal phenotype when high levels are reached. The use of seed-specific promoters alleviates these undesired properties, with the expected increase in threonine and lysine synthesis, but in the case of lysine without accumulation in the seed. Control of lysine catabolism may play an important role in the future to solve this problem. The availability of the first cloned corresponding plant genes also opens the way to similar types of metabolic engineering. Coding sequences of insensitive forms of the two key enzymes will be fused to appropriate promoters depending on the organ and crop species that are targeted for high lysine content. This approach to nutritional improvement of crop plants could also be combined with the expression of a lysine-rich storage protein which would provide an increased local demand in this amino acid.

From a fundamental point of view, the cloning and characterization of genes coding for enzymes involved in primary amino acid metabolism in plants brings invaluable information on the developmental and tissue-specific expression of these genes and their integration in the complex network of amino acid metabolism. Coding regions will be also exploited for the spatial and temporal suppression of gene activity based on antisense constructs. Reporter genes to be fused to promoter sequences will surely be used for transient and stable transformation experiments in order to dissect the regulation of the key aspartate family genes.

Putative regulatory elements identified in the genomic sequences will be tested for functionality and in conjunction with varied physiological conditions to uncover modes of regulation which have not yet been revealed in such pathways.

References

Boyes, C.J. & Vasil, I.K. (1987). *In vitro* selection for tolerance to *S*-(2-aminoethyl)-L-cysteine and overproduction of lysine by embryogenic calli and regenerated plants of *Pennisetum americanum* (L.) K. Schum. *Plant Science,* **50**, 195–203.

Bright, S.W.J., Featherstone, L.C. & Miflin, B.J. (1979a). Lysine metabolism in a barley mutant resistant to *S*-(2-aminoethyl)-L-cysteine. *Planta,* **146**, 629–33.

Bright, S.W.J., Kueh, J.S.H., Franklin, J., Rognes, S.E. & Miflin, B.J. (1982a). Two genes for threonine accumulation in barley. *Nature, London,* **299**, 278–9.

Bright, S.W.J., Miflin, B.J. & Rognes, S.E. (1982b). Threonine accumulation in the seeds of a barley mutant with an altered aspartate kinase. *Biochemical Genetics,* **20**, 229–43.

Bright, S.W.J., Norbury, P.B. & Miflin, B.J. (1979b). Isolation of a recessive barley mutant resistant to *S*-(2-aminoethyl)-L-cysteine. *Theoretical and Applied Genetics,* **55**, 1–4.

Cattoir-Reynaerts, A., Degryse, E. & Jacobs, M. (1981). Selection and analysis of mutants overproducing amino acids of the aspartate family in barley, *Arabidopsis* and carrot. In *Induced Mutations – A Tool in Plant Research* (Proc. Symp. Vienna), pp. 353–361. Vienna: IEAE.

Cattoir-Reynaerts, A., Degryse, E., Verbruggen, I. & Jacobs, M. (1983). Selection and characterization of carrot embryoid cultures resistant to inhibition by lysine plus threonine. *Biochemie und Physiologie der Pflanzen,* **178**, 81–90.

Colau, D. (1988). Transfert et expression, dans des cellules de *Nicotiana*, de genes étrangers codant pour des acides amines impliqués dans la biosynthèse d'acides aminés derivés de l'acide aspartique. PhD thesis, Vrije Universiteit Brussel.

Datta, P. (1969). Regulation of branched biosynthetic pathways in bacteria. *Science,* **165**, 556–62.

Diedrick, T.J., Frisch, D.A. & Gengenbach, B.G. (1990). Tissue culture isolation of a second locus for increased threonine accumulation in maize. *Theoretical and Applied Genetics,* **79**, 209–15.

Dotson, S.B., Frisch, D.A., Somers, D.A. & Gengenbach, B.G. (1990). Lysine-insensitive aspartate kinase in two threonine-overproducing mutants of maize. *Planta,* **182**, 546–52.

Frankard, V., Ghislain, M. & Jacobs, M. (1992). Two feedback-insensitive enzymes of the aspartate pathway in *Nicotiana sylvestris*. *Plant Physiology,* **99**, 1285–93.

Frankard, V., Ghislain, M., Negrutiu, I. & Jacobs, M. (1991). High threonine producer mutant in *Nicotiana sylvestris* (Spegg. and Comes). *Theoretical and Applied Genetics,* **82**, 273–82.

Frisch, D.A., Tommey, A.M., Gengenbach, B.G. & Somers, D.A. (1991). Direct genetic selection of a maize cDNA for dihydro-dipicolinate synthase in an *Escherichia coli dapA*-auxotroph. *Molecular General Genetics,* **228**, 287–93.

Gebhardt, J.S., Weiseman, J.M. & Matthews, B.F. (1993). Molecular analysis of the aspartate kinase-homoserine dehydrogenase gene family in soybean. *Plant Physiology* (Suppl), 377.

Gengenbach, B.G. (1984). Tissue culture and related approaches for grain quality improvement. In *Applications of Genetic Engineering to Crop Improvement*, ed. G.B. Collins & J.F.Petolino, pp. 211–54. Dordrecht: Nijhoff-Junk.

Ghislain, M., Frankard, V., Van Den Bossche, D., Matthews, B. & Jacobs, M. (1994). Molecular analysis of the aspartate kinase-homoserine dehydrogenase gene from *Arabidopsis thaliana*. *Plant Molecular Biology* **24**, 835–51.

Glassman, K.F. (1992). A molecular approach to elevating free lysine in plants. In *Biosynthesis and Molecular Regulation of Amino Acids in Plants*, ed. B.K. Singh, H.E. Flores & J.C.Shannon, pp. 217–28. Rockville: American Society of Plant Physiologists.

Hibberd, K.A. & Green, C.E. (1982). Inheritance and expression of lysine plus threonine resistance selected in maize tissue culture. *Proceedings of National Academy of Sciences USA*, **79**, 559–63.

Hibberd, K.A., Walter, T., Green, C.E. & Gengenbach, B.G. (1980). Selection and characterization of a feedback-insensitive tissue culture of maize. *Planta*, **148**, 183–7.

Jacobsen, E. (1986). Isolation, characterization and regeneration of an *S*-(2-aminoethyl)-L-cysteine resistant cell line of dihaploid potato. *Journal of Plant Physiology*, **123**, 307–15.

Kaneko, T., Hashimoto, T., Kumpaisal, R. & Yamada, Y. (1990). Molecular cloning of wheat dihydrodipicolinate synthase. *Journal of Biology and Chemistry*, **265**, 17451–5.

Karchi, H., Shaul, O. & Galili, G. (1993). Seed-specific expression of a bacterial desensitized aspartate kinase increases the production of seed threonine and methionine in transgenic tobacco. *The Plant Journal*, **3**, 721–7.

Karchi, H., Shaul, O. & Galili, G. (1994). Lysine synthesis and catabolism are coordinately regulated during tobacco seed development. *Proceedings of the National Academy of Sciences, USA*, **91**, 2577–81.

Kumpaisal, R., Hashimoto, T. & Yamada, Y. (1988). Selection of *S*-(2-aminoethyl)-L-cysteine resistant wheat cell cultures. *Journal of Plant Physiology*, **133**, 608–14.

Last, R.L. (1992). The genetics of nitrogen assimilation and amino acid biosynthesis in flowering plants: progress and prospects. *International Review of Cytology*, **143**, 297–330.

Matthews, B.F., Shye, S.C.H. & Widholm, J.M. (1980). Mechanism of resistance of a selected carrot cell suspension to *S*-(2-aminoethyl)-L-cysteine. *Zeitschrift für Pflanzenphysiologie*, **96**, 453–63.

Miao, S., Duncan, D.R. & Widholm, J. (1988). Selection of regenerable maize callus cultures resistant to 5-methyl-DL-tryptophan, *S*-(2-aminoethyl)-L-cysteine and high levels of L-lysine plus L-threonine. *Plant Cell, Tissue and Organ Culture*, **14**, 3–14.

Muehlbauer, G.J., Somers, D.A., Matthews, B.F. & Gengenbach, B.G. (1993). Molecular genetics of maize aspartate kinase-homoserine dehydrogenase. *Plant Physiology (Suppl.)*, 375.

Negrutiu, I., Cattoir-Reynaerts, A., Verbruggen, I. & Jacobs, M. (1984). Lysine overproducer mutants with an altered dihydrodipicolinate synthase from protoplast culture of *Nicotiana sylvestris* (Spegazzini and Comes). *Theoretical and Applied Genetics*, **68**, 11–20.

Perl, A., Shaul, O. & Galili, G. (1992). Regulation of lysine synthesis in transgenic potato plants expressing a bacterial dihydrodipicolinate synthase in their chloroplasts. *Plant Molecular Biology*, **19**, 815–23.

Rognes, S.E., Wallsgrove, R.M., Kueh, J.S.H. & Bright, S.W.J. (1986). Effect of exogenous amino acids on growth and activity of four aspartate pathway enzymes in barley. *Plant Science*, **43**, 45–50.

Schaeffer, G.W. & Sharpe, F.T. (1981). Lysine in seed protein from S-aminoethyl-L-cysteine resistant anther-derived tissue cultures of rice. *In Vitro*, **17**, 345–52.

Schaeffer, G.W. & Sharpe, F.T. (1987). Increased lysine and seed storage protein in rice plants recovered from calli selected with inhibitory levels of lysine plus threonine and S-aminoethyl-L-cysteine. *Plant Physiology*, **84**, 509–15.

Schaeffer, G.W. & Sharpe, F.T. (1990). Modification of amino acid composition of endosperm proteins from *in vitro*-selected high lysine mutants in rice. *Theoretical and Applied Genetics*, **80**, 841–6.

Shaul, O. & Galili, G. (1992a). Increased lysine synthesis in tobacco plants that express high levels of bacterial dihydrodipicolinate synthase in their chloroplasts. *The Plant Journal*, **2**, 203–9.

Shaul, O. & Galili, G. (1992b). Threonine overproduction in transgenic tobacco plants expressing a desensitized aspartate kinase of *Escherichia coli*. *Plant Physiology*, **100**, 1157–63.

Silk, G.W. & Matthews, B.F. (1993). Cloning and expression of the soybean lysine pathway gene *dapA*. *Plant Physiology (Suppl.)*, 83.

Thompson, G.A., Datko, A.H., Mudd, S.H. & Giovanelli, J. (1982). Methionine biosynthesis in *Lemna*. Studies on the regulation of cystathionine-γ-synthase, o-phosphohomoserine sulfhydrylase, and o-acetylserine sulfhydrylase. *Plant Physiology*, **69**, 1077–83.

Vernaillen, S., Van Ghelue, M., Verbruggen, I. & Jacobs, M. (1985). Characterisation of mutants in *Arabidopsis thaliana* (L.) Heynh. resistant to analogues and inhibitory concentrations of amino acids derived from aspartate. *Arabidopsis Information Service*, **22**, 13–22.

Wallsgrove, R.M., Lea, P.J. & Miflin, B.J. (1983). Intracellular localisation of aspartate kinase and the enzymes of threonine and methionine biosynthesis in green leaves. *Plant Physiology*, **71**, 780–4.

Weisemann, J. & Matthews, B.F. (1993). Identification and expression of a cDNA from *Daucus carota* encoding a bifunctional aspartokinase-homoserine dehydrogenase. *Plant Molecular Biology*, **22**, 301–12.

P. JOHN

Oxidation of 1-aminocyclopropane-1-carboxylic acid (ACC) in the generation of ethylene by plants

Introduction

Ethylene is the plant growth regulator that controls ripening and senescence in plants. It is produced from methionine via the formation of
S-adenosyl-L-methionine, which, in turn, forms the non-protein amino
acid, 1-aminocyclopropane-1-carboxylic acid (ACC) (Fig. 1). When ACC
was first recognized as the immediate substrate for ethylene formation, it
became apparent that all kinds of plant tissues had an active and constitutive enzyme capable of catalysing the conversion of ACC to ethylene, as
high rates of ethylene production were observed when plant tissues were
supplied with ACC (Cameron *et al.*, 1979). The rate of ethylene production
when tissues were fed ACC was always far greater than when they were fed
methionine under the same conditions (Yang & Hoffman, 1984). During
climacteric fruit ripening and flower senescence an initially low ACC oxidase activity increased in line with the rise of ACC synthesis and ethylene
production (Yang & Hoffman, 1984). However, despite the substantial
ACC oxidase activity that was readily observed with plant tissues, *in vitro*
activity could not be detected. When tissues were homogenized, methionine was readily converted enzymatically to ACC, but there was no ethylene
formation, and ACC supplied to tissue homogenates was not converted
to ethylene. These observations led to the notion that ethylene formation
required membrane integrity (Yang & Hoffman, 1984). This view was supported by the sensitivity of ACC oxidase to osmotic shock, detergents and
protonophores, and by the discontinuities in the Arrhenius plots for ethylene production (see Yang & Hoffman, 1984; Kende, 1993). Uncertainty
over the nature of the enzyme responsible for the generation of ethylene
from ACC was reflected in the relatively uninformative trivial name that
was used: ethylene-forming enzyme (EFE).

ACC oxidase activity requires Fe^2 and ascorbate *in vitro*

For the ACC oxidase the 'dark ages' were brought to a close by the pioneering genetic transformation work on tomato by Hamilton, Lycett and

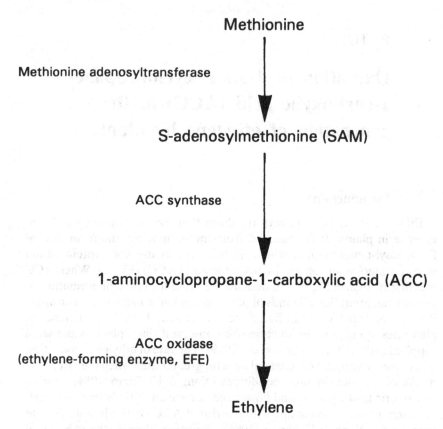

Fig. 1. The pathway for ethylene synthesis from methionine in plants.

Grierson (1990). They reported that both ethylene production by the fruits and ACC oxidase activity measured in leaf discs were reduced in plants that had been transformed by a 1.1 kb fragment of a cDNA, designated pTOM13, inserted in an antisense orientation. The extent to which ethylene synthesis and ACC oxidase activity were reduced in the transformed plants depended on the gene dosage: in plants inheriting two antisense genes, ethylene synthesis was inhibited by 97% and ACC oxidase by 93%. It was established that pTOM13 encoded for a protein of 35 kDa on an SDS gel, and that it was one of a limited group of proteins which increased in abundance during fruit ripening. However, conclusive identification of the pTOM13 protein as the ACC oxidase had to await development of an assay for the *in vitro* enzyme.

Hamilton *et al.* (1990) reported that the predicted amino acid sequence of the pTOM13 polypeptide showed a 33% identity and a 58% similarity to flavanone-3-hydroxylase from *Antirrhinum majus*. This is a soluble enzyme responsible for the hydroxylation of 2S flavanones to form 2R,3R dihydroflavonols in the biosynthetic pathway to anthocyanidins. At this time, ACC oxidase was still being sought in the membrane fractions of tissue homogenates, and there was no obvious enzymatic relationship between ACC oxidase and flavanone-3-hydroxylase. However, Ververidis and John (1991), noting that flavanone-3-hydroxylase required certain conditions to maintain enzyme stability *in vitro* (Britsch & Grisebach, 1986) adopted these conditions for the extraction and assay of ACC oxidase. As a result, Ververidis & John (1991) obtained complete recovery of activity from ripening melons, and it was in the soluble fraction that the ACC oxidase was found; the enzyme could function without the membrane fraction. They showed that ACC oxidase activity required ascorbate and Fe^{2+}, but unlike flavanone-3-hydroxylase, ACC oxidase did not require 2-oxoglutarate (Smith, Ververidis & John, 1992). When functional expression of ACC oxidase was obtained in yeast cells that had been transformed with a pTOM13 gene (Hamilton, Bouzayen & Grierson, 1991), the requirement of the enzyme for ascorbate and Fe^{2+} was immediately apparent as a stimulatory effect on the ACC oxidase activity observed when the transformed yeast cells were provided with ACC.

Since the initial discovery (Ververidis & John, 1991) of the conditions for the assay of ACC oxidase *in vitro*, progress has been rapid and spectacular. Stabilization of enzyme activity during purification has been enhanced by the chelation of Fe^{2+} with 1,10-phenanthroline (Dupille *et al.*, 1993), and by the use of 30% glycerol in the extraction media (Dong, Fernández-Maculet & Yang, 1992; Smith & John, 1993a,b). ACC oxidases have been characterized, either after complete or partial purification, from fruits of melon (Smith *et al.*, 1992), avocado (McGarvey & Christoffersen, 1992; Christoffersen, McGarvey & Savarese, 1993) and apple (Kuai & Dilley, 1992; Dong *et al.*, 1992; Dupille *et al.*, 1993; Dilley *et al.*, 1993). The different ACC oxidases appear to resemble one another in a number of respects. The pH optimum is found to be around pH 7.5 (Smith *et al.*, 1992; McGarvey & Christoffersen, 1992). The protein is active as a monomer of about 40 kDa as determined by gel filtration (Smith *et al.*, 1992; Dong *et al.*, 1992; Dupille *et al.*, 1993) and about 35 kDa as determined by SDS PAGE (Dong *et al.*, 1992).

ACC oxidase has an absolute requirement for both Fe^{2+} and ascorbate. Presumably the Fe^{2+} requirement is created because Fe^{2+} is easily lost from the enzyme during its purification. In the 2-oxoacid-dependent dioxy-genases (Prescott, 1993), which are related by sequence homology to the ACC oxidase, ascorbate is a cofactor, required as a reducing agent. By contrast, in ACC oxidase, ascorbate appears to be an essential co-substrate (Dong et al., 1992; see also Smith et al. 1992), being oxidized to dehydroascorbate stoichiometrically during the oxidation of ACC to ethylene (Dong et al., 1992). Thus it appears that in the reaction catalysed by the ACC oxidase, one of the oxygen atoms of dioxygen oxidizes ascorbate and the other atom oxidises ACC (Fig. 2). In this respect, ascorbate plays a role in the reaction mechanism of the ACC oxidase that is normally played by 2-oxoglutarate in the structurally related 2-oxoglutarate-dependent dioxy-genases (Prescott, 1993).

The fact that in vitro ACC oxidase produces virtually no ethylene in the absence of ascorbate raises the question of how ACC oxidase can operate in transformed yeast, as yeast apparently lacks ascorbate (see, for example, Moser & Bendich, 1990; Bruins, Scharloo & Thorig, 1991). Hamilton et al. (1991) observed that, with suspensions of yeast cells transformed with the tomato ACC oxidase, the maximum rate of ethylene production was obtained when 50 mM ascorbate was added together with the ACC. But when ascorbate was not added, 24% of this maximum rate (27% in the presence of added Fe^{2+}) could still be observed (Hamilton et al., 1991). Similarly, Wilson et al. (1993), working with yeast transformed with the apple ACC oxidase, found that about 50% of the maximum rate of ethylene production could be observed when ascorbate was omitted from the assay medium. Possibly, when ascorbate is unavailable, as in yeast, it is replaced by another metabolic reductant, although none of the other reductants that have been tested are able to support the ACC oxidase activity in vitro (Smith et al., 1992).

Carbon dioxide as a cofactor

One of the most important recent findings made with ACC oxidase has been its requirement for carbon dioxide as an essential cofactor.

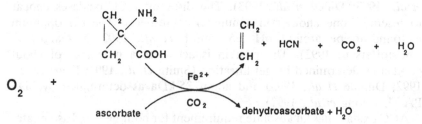

Fig. 2. The reaction catalysed in vitro by the ACC oxidase of plants.

From work with the enzyme from apple (Dong *et al.*, 1982; Dilley *et al.*, 1993) and melon (Smith & John, 1993*a,b*) it is now clear that in the absence of carbon dioxide, no ACC activity is detectable (Dong *et al.*, 1992). Most investigators of the *in vivo* and *in vitro* ACC oxidase have been unaware of this requirement, and consequently the activities reported have been depended upon ambient levels of carbon dioxide. Compared to the activity observed with ambient levels of carbon dioxide, addition of about 5% carbon dioxide to the atmosphere above the assay medium, or 25 mM $NaHCO_3$ in the assay medium, will stimulate *in vitro* ACC oxidase activity 10- to 20-fold (Dong *et al.*, 1992; Smith & John 1993*a,b*). In the case of the ACC oxidase from melon (Smith & John, 1993*a,b*), the carbon dioxide requirement made itself known when there was a loss of activity during purification attributable to the loss of a low molecular weight factor. This factor could be replaced by bicarbonate or by carbon dioxide, but it appears that carbon dioxide rather than the bicarbonate anion is the active species (Smith & John, 1993*a,b*). In its activation by carbon dioxide, ACC oxidase resembles Rubisco, and it remains to be seen whether the mechanism of activation in ACC oxidase involves a carbamylation of a lysine residue as it does in Rubisco (Lorimer, 1983).

Even when supplied with saturating levels of carbon dioxide, the rate at which ACC oxidase turns over its substrates *in vitro* is very low. Thus the enzyme purified from apple is reported (Dong *et al.*, 1992; Yang *et al.*, 1993) to have a specific activity of 20 nmol mg^{-1} protein min^{-1}. This can be expressed as a turnover number of 0.8 mol substrate mol^{-1} enzyme min^{-1}, which compares with a turnover number of 85 mol substrate mol^{-1} enzyme min^{-1} calculated for the ACC synthase purified from apple (Yip, Dong & Yang, 1991). A similarly low turnover number has been observed for the tomato ACC oxidase expressed in *Escherichia coli* and purified as a fusion protein (Z.H. Zhang & J. Smith, unpublished observations), while the ACC synthase purified from tomato has a turnover number of 370 mol substrate mol^{-1} enzyme min^{-1} (Bleeker *et al.*, 1986). Thus, although optimum *in vitro* conditions appear to have been found for the ACC oxidase, each enzyme molecule is turning over less than one substrate molecule per min.

Catalytic inactivation

The ACC oxidase activity is generally assayed by measuring the ethylene accumulated over a period of 10–20 min. If the incubation is continued for longer, activity declines and the time-course becomes significantly non-linear (Smith *et al.*, 1992; Dilley *et al.*, 1993). Smith and John (1993*a,b*) have shown that this activity decline is attributable to the inactivation of the enzyme during catalysis. Preincubation of the melon enzyme in the presence of either ACC, ascorbate or Fe^{2+} added singly does not significantly affect enzyme activity, but when the enzyme is preincubated in the presence of all three factors it becomes inactivated (Smith & John 1993*a,b*). Passage of the inactivated enzyme through a column of Sephadex G-25 does not lead to a reactivation, and inactivation is independent of the number of catalytic cycles (Smith & John, unpublished data). Therefore end-product accumulation does not appear to be responsible. The mechanism responsible for the catalytic inactivation of ACC oxidase is not yet known, but it is notable that it is also a feature of ACC synthase (see Kende, 1993), the preceding enzyme in the biosynthetic pathway to ethylene.

Acknowledgements

Research in the author's laboratory is funded by the EC ECLAIR programme and Zeneca-AFRC.

References

Bleeker, A.B., Kenyon, W.H., Somerville, S.C. & Kende, H. (1986). Use of monoclonal antibodies in the purification and characterization of 1-aminocyclopropane-1-carboxylate synthase, and enzyme in ethylene biosynthesis. *Proceedings of the National Academy of Sciences, USA*, **83**, 7755–9.

Britsch, L. & Grisebach, H. (1986). Purification and characterization of (2S) -flavanone 3-hydroxylase from *Petunia hybrida*. *European Journal of Biochemistry*, **156**, 569–77.

Bruins, B.G., Scharloo, W. & Thorig, G.E.W. (1991). Dietary ascorbic acid, pyridoxine and riboflavin reduce the light sensitivity of larvae and pupae of *Drosophila melanogaster*. *Insect Biochemistry*, **21**, 541–4.

Cameron, A.C., Fenton, C.A.L, Yu, Y., Adams, D.O. & Yang, S.F. (1979). Increased production of ethylene by plant tissues with 1-aminocyclopropane-1-carboxylic acid. *HortScience*, **14**, 178–80.

Christoffersen, R.E., McGarvey, D.J. & Savarese, P. (1993). Biochemical and molecular characterization of ethylene forming enzyme from avocado. In *Cellular and Molecular Aspects of the Plant Hormone Ethylene*, ed. J.-C. Pech, A. Latché & C. Balagué, pp. 65–70. Dordrecht: Kluwer.

Dilley, D.R., Kuai, J., Poneleit, L., Zhu, Y., Pekker, Y., Wilson, I.D., Burmeister, D.M., Gran, C. & Bowers, A. (1993). Purification and characterization of ACC oxidase and its expression during ripening in apple fruit. In *Cellular and Molecular Aspects of the*

Plant Hormone Ethylene, ed. J.-C. Pech, A. Latché & C. Balagué, pp. 46–52. Dordrecht: Kluwer.

Dong, J.-G., Fernández-Maculet, J.C. & Yang, S.F. (1992). *In vitro* stabilization and purification of 1-aminocyclopropane-1-oxidase from ripe apple fruit. *Proceedings of the National Academy of Sciences, USA*, **89**, 9789–93.

Dupille, E., Rombaldi, C., Lelièvre, J.-M., Cleyet-Marel, J.C., Pech, J.-C. & Latché, A. (1993). Purification, properties and partial amino-acid sequence of 1-aminocyclopropane-1-carboxylic acid oxidase from apple fruits. *Planta*, **190**, 65–70.

Hamilton, A.J., Bouzayen, M., & Grierson, D. (1991). Identification of a tomato gene for the ethylene-forming enzyme by expression in yeast. *Proceedings of the National Academy of Sciences, USA*, **88**, 7434–7.

Hamilton, A.J., Lycett, G.W. & Grierson, D. (1990). Antisense gene that inhibits synthesis of the hormone ethylene in transgenic plants. *Nature, London*, **346**, 284–7.

Kende, H. (1993). Ethylene biosynthesis. *Annual Review of Plant Physiology and Plant Molecular Biology*, **44**, 283–307.

Kuai, J. & Dilley, D.R. (1992). Extraction, partial purification and characterization of 1-aminocyclopropane-1-carboxylic acid oxidase from apple fruit. *Postharvest Biology and Technology*, **1**, 203–11.

Lorimer, G.H. (1983). Carbon dioxide and carbamate formation: the makings of a biochemical control system. *Trends in Biochemical Sciences*, **8**, 65–8.

McGarvey, D.G. & Christoffersen, R.E. (1992). Characterization and kinetic parameters of ethylene-forming enzyme from avocado. *Journal of Biological Chemistry*, **267**, 5964–7.

Moser, U. & Bendich, A. (1990). Vitamin C. In *Handbook of Vitamins* 2nd edition, ed. L.J. Machlin, pp. 195–232. New York: Marcel Decker.

Prescott, A.G. (1993). A dilemma of dioxygenases (or where biochemistry and molecular biology fail to meet). *Journal of Experimental Botany*, **44**, 849–61.

Smith, J.J. & John, P. (1993*a*). Activation of 1-aminocyclopropane-1-carboxylate oxidase by bicarbonate/carbon dioxide. *Phytochemistry*, **32**, 1381–6.

Smith, J.J. & John, P. (1993*b*). Maximising the activity of the ethylene-forming enzyme. In *Cellular and Molecular Aspects of the Plant Hormone Ethylene*, ed. J.-C. Pech, A. Latché & C. Balagué, pp. 33–8. Dordrecht: Kluwer.

Smith, J.J., Ververidis, P. & John, P. (1992). Characterisation of the ethylene-forming enzyme partially purified from melon. *Phytochemistry*, **31**, 1485–94.

Ververidis, P. & John, P. (1991). Complete recovery *in vitro* of ethylene forming enzyme activity, *Phytochemistry*, **30**, 725–7.

Wilson, I.D., Zhu, Y., Burmeister, D.M. & Dilley, D.R. (1993). Apple ripening-related cDNA clone pAP4 confers ethylene-forming ability in transformed *Saccharomyces cerevisiae*. *Plant Physiology*, **102**, 783–8.

Yang, S.F., Dong, J.G., Fernández-Maculet, J.C. & Olsen, D.C. (1993). Apple ACC oxidase: purification and characterization of the enzyme and cloning of its cDNA. In *Cellular and Molecular Aspects of the Plant Hormone Ethylene*, ed. J.-C. Pech, A. Latché & C. Balagué, pp 59–64. Dordrecht: Kluwer.

Yang, S.F. & Hoffman, N.E. (1984). Ethylene biosynthesis and its regulation in higher plants. *Annual Review of Plant Physiology*, **35**, 155–89.

Yip, W.-K., Dong, J.-G. & Yang, S.F. (1991). Purification and characterisation of 1-aminocyclopropane-1-carboxylate synthase from apple. *Plant Physiology*, **95**, 251–7.

BIJAY K. SINGH, IWONA SZAMOSI and
DALE L. SHANER

Regulation of carbon flow through the branched chain amino acid biosynthetic pathway

The branched chain amino acid biosynthetic pathway has received considerable attention in recent years because different chemical classes of highly successful commercial herbicides kill plants by inhibiting this pathway. This discovery has led to identification and design of other inhibitors of this pathway that are also herbicidal. Our recent studies on the mode of action of imidazolinone herbicides have provided clues that may help clarify the role of 2-ketobutyrate and 2-aminobutyrate in the phytotoxic effects of these herbicides. These studies have also provided insight into understanding the regulation of carbon flow through the branched chain amino acid biosynthetic pathway in plants. We have also identified a new form of threonine dehydratase (TD; EC 4.2.1.16) that may have a crucial function in nitrogen metabolism in senescing leaves.

Accumulation of 2-KB/2-AB vs phytotoxicity of AHAS inhibitors

The imidazolinone and sulfonylurea families of highly successful commercial herbicides kill plants by inhibiting acetohydroxyacid synthase (AHAS; EC 4.1.3.18), the first common enzyme in the pathways leading to the biosynthesis of valine, leucine and isoleucine. These extremely potent herbicides kill plants at application rates of grams per hectare. The high potency of AHAS inhibiting herbicides is of great interest because inhibitors of other enzymes in the branched chain amino acid pathway require much higher rates to kill plants (Wittenbach, Aulabaugh & Schloss, 1991; Shaner & Singh, 1992). Herbicidal effects of the AHAS inhibitors may result from any one or a combination of the following mechanisms: depletion of the end products; depletion of intermediates of the pathway for some critical processes; or a build-up of a toxic substrate.

Inhibition of AHAS in *Salmonella typhimurium* leads to accumulation of 2-ketobutyrate (2-KB), one of the substrates of AHAS (LaRossa, Van Dyke & Smulski, 1987). High levels of 2-KB are toxic to *S.typhimurium*. Similarly, chlorsulfuron treatment led to the accumulation of 2-aminobutyrate (2-AB), the transamination product of 2-KB, in Lemna minor (Rhodes *et al.*, 1987). 2-AB has also been shown to disrupt cell division in *Allium* (Langzagorta de la Torre & Aller, 1988) and *Hordeum* root tips (Reid, Field & Pitman, 1985). These observations led to the hypothesis that inhibition of AHAS causes accumulation of 2-KB and 2-AB and this build-up is what kills the microorganism or plant (LaRossa & Van Dyk, 1987; LaRossa *et al.*, 1987, 1990; Rhodes *et al.*, 1987; Schloss, 1989; Schloss & Aulabaugh, 1989; Van Dyk & La Rossa, 1986; 1987; Van Dyk *et al.*, 1987). It was also postulated that other inhibitors in the pathway, such as Hoe 704 which inhibits ketoacidreductoisomerase (KARI; EC 1.1.1.86; Schultz *et al.*, 1988), are not as potent as the AHAS inhibitors because they do not cause the accumulation of 2-KB or 2-AB or another toxic substrate. If these assumptions are true then treatments that would decrease the accumulation of 2-KB should alleviate toxicity of AHAS inhibitors. On the other hand, treatments that would cause accumulation of 2-KB/2-AB in the plant should result in plant death.

Decrease in the level of 2-KB/2-AB

Imazaquin inhibits leaf elongation of corn seedlings and causes accumulation of both 2-KB and 2-AB (Fig.1). Interestingly, accumulation of 2-AB was about 100-fold greater than accumulation of 2-KB suggesting that the two compounds stay in equilibrium *in vivo*, the equilibrium being in favour of 2-AB. Since 2-AB is the major component of the two, and the accumulation trend for the two compounds was the same, only the data for 2-AB will be discussed here.

LaRossa *et al.* (1987) showed that isoleucine plus valine reverse the toxicity of sulfonylureas in the wild type *S.typhimurium* and suggested that this reversal was due to isoleucine preventing the production of 2-KB by inhibiting TD. Such a reversal by isoleucine was not observed when mutants expressing an isoleucine-insensitive form of TD were grown in the presence of sulfometuron methyl.

Since TD in plants is also feedback regulated by isoleucine, supplementation with isoleucine should prevent the herbicidal effects of the AHAS inhibiting herbicides. In order to test this hypothesis, intact corn plants were treated with imazaquin, valine, leucine and isoleucine, either singly or in various combinations via the root system. Leaf

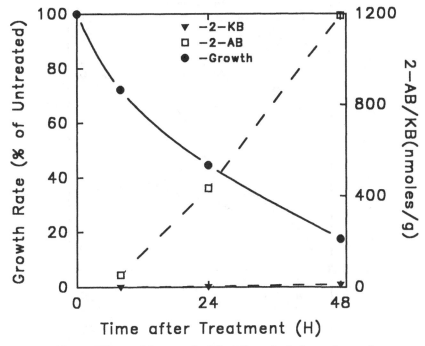

Fig. 1. Effects of imazaquin (10 μM) on leaf elongation and accumulation of 2-ketobutyrate and 2-aminobutyrate in corn seedlings. Growth rates were determined by measuring the rate of elongation of the fourth leaf.

elongation of corn seedlings was inhibited by imazaquin (Fig. 2A). Ile, val+ile and val+leu+ile did not have a significant effect on leaf elongation. Valine alone inhibited growth but not to the same extent as the inhibition observed with imazaquin. Isoleucine or valine alone did not reverse the herbicidal effects of imazaquin (Fig. 2B), however, val+ile did cause a significant reversal of growth inhibition by imazaquin. Nearly complete reversal of the growth inhibitory effects of imazaquin were obtained by supplementing the plant with a combination of val+leu+ile.

Imazaquin treatment caused about 250-fold increase in the concentration of 2-AB (Table 1). Ile supplementation with imazaquin reduced 2-AB levels to about 10% of the level present in the imazaquin-treated plants, yet ile did not protect plants from the inhibitory effects of imazaquin. Interestingly, a 7-fold higher level of 2-AB was observed in val-treated plants compared to the imazaquin-treated plants, although

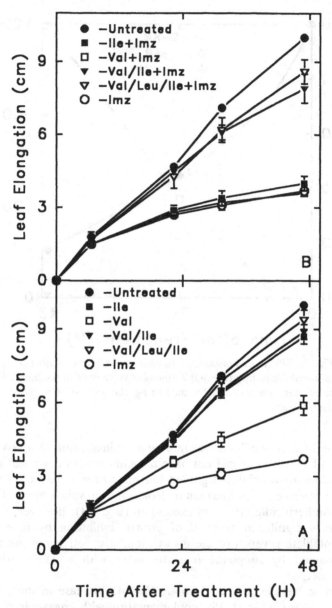

Fig 2. Effects of imazaquin in various combinations with ile, val, and leu on growth rate in hydroponically grown, 10-day-old corn seedlings. Treatments are as indicated in the Figure.

Table 1. *Amino acid composition (nmoles/g fresh weight) of corn plants after 48 h of individual or combination treatment with imazaquin, val, leu and isoleucine as shown in the Table*

	Control	Imz	Ile	Val	Val Ile	Val Leu Ile	Ile Imz	Val Imz	Val Ile Imz	Val Leu Ile Imz
Asp	1105	2655	1141	1292	1051	1029	2804	2348	1678	1023
Thr+Ser	1541	5979	2294	3109	2533	2146	5958	4835	5103	2152
Glu	1201	3467	1197	1563	1154	1159	3796	2758	1991	1207
Pro	20	137	29	49	13	20	120	119	55	23
Gly	91	1843	459	1319	595	436	2068	1846	1362	494
Ala	183	4700	491	3120	637	477	5207	3522	2454	541
2-AB	0.6	148	0.5	1036	0.5	0.6	14	1218	0.6	0.5
Val	66	21	461	2773	2153	1299	75	5266	2307	1524
Met	0.8	188	22	64	15	18	109	187	34	21
Ile	23	189	2979	106	3188	2294	3612	7	2502	2439
Leu	39	21	71	1247	62	3023	42	2908	26	3512
Tyr	73	535	127	241	125	152	609	464	198	169
Phe	12	120	38	51	22	49	181	119	55	53
His	38	273	42	112	60	69	192	240	90	61
Lys	16	258	27	45	37	37	226	177	41	44
Arg	96	1125	192	625	207	202	855	1068	297	224
Total	4506	21659	9569	16750	11850	12411	25869	27083	18191	13487

val did not reduce growth to the same extent as did imazaquin. These results showed there was no correlation between the levels of 2-AB in plant and growth inhibition caused by the inhibition of AHAS either by valine or imazaquin.

Increase in the level of 2-KB/2-AB

If 2-KB/2-AB is toxic to plants then feeding 2-KB/2-AB to a plant should cause accumulation of these compounds and result in growth inhibition and eventually plant death. Since 2-KB and 2-AB are in equilibrium *in vivo*, feeding either of these compounds should have the same effect. 2-KB is not readily taken up by plant roots (unpublished data), therefore different concentrations of 2-AB or imazaquin were fed through the root system. Root-applied imazaquin inhibited growth in a dose dependent manner (Fig.3A). However, up to 2 mM 2-AB did not have a significant effect on leaf elongation of corn plants (Fig. 3B). Interestingly, the lowest concentration of 2-AB (0.5 mM) caused a nearly two-fold greater accumulation of 2-AB within the shoot meristem region than the highest concentration of imazaquin (10 μM) (Table 2). However, there was no significant effect of this level of 2-AB treatment on plant growth. Feeding 2 mM 2-AB resulted in a build up of 2-AB in vivo to 1978 nmol/g fresh weight, however, even this high internal concentration of 2-AB did not significantly affect the plant growth. Similarly, there was no correlation between growth inhibition and 2-KB accumulation (data not shown). 2-AB feeding did result in accumulation of isoleucine (Table 2) which shows that 2-AB taken up by the root system must be transported to the chloroplast where it is converted to isoleucine by the enzymes of this pathway. Therefore, externally fed 2-AB is transported to the chloroplast where it is supposed to be produced in the imidazolinone-treated plants. This result rules out the possibility that exogenously supplied 2-AB accumulates in different plant parts and in different sub-cellular fraction than the location where 2-AB would be expected to accumulate in imazaquin treated plants.

Accumulation of 2-KB/2-AB due to inhibition of KARI

It has been suggested that KARI is not a good herbicide target site because inhibition of this enzyme does not cause accumulation of a toxic substrate. We postulated that blockage of the branched chain amino acid biosynthetic pathway due to inhibition of KARI might cause accumulation of other intermediates of the pathway besides acetolactate and acetohydroxybutyrate. In order to test this hypothesis, corn seedlings were treated with imazaquin or Hoe704, an inhibitor of KARI. Both herbicides caused accumulation of 2-KB and 2-AB (only data for 2-AB presented in Fig.4). These data do not support the proposal that the higher potency of AHAS inhibitors compared to KARI inhibitors is due to differential effects on the levels 2-KB/2-AB in plant tissue. Therefore, the results from the three sets of experiments described here demonstrate that the initial phytotoxic effects of AHAS-inhibiting herbicides are not due to accumulation of 2-KB or 2-AB.

Interaction between the pools of the branched chain amino acids

The branched chain amino acid biosynthetic pathway is regulated at a number of steps: (i) valine and leucine inhibit acetohydroxyacid syn-

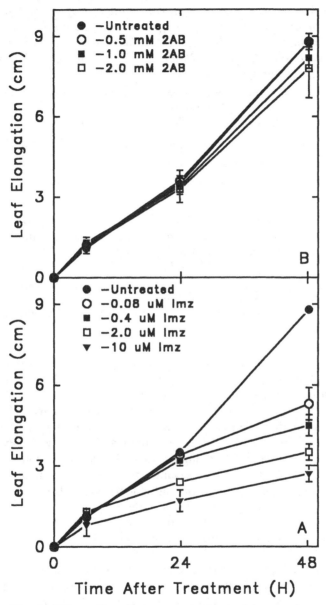

Fig. 3. Effects of various concentrations of imazaquin or 2-AB on elongation of the fourth leaf of hydroponically grown, 10-day-old corn seedlings.

Table 2. *Effects of different concentrations of imazaquin and 2-aminobutyrate on leaf elongation rates, 2-aminobutyrate accumulation and free branched chain amino acid levels in hydroponically grown, 10-day-old corn seedlings 24 hours after initiation of treatment*

Treatment	Concentration	Growth rate[a] (mm/h)	2-AB	Ile	Val	Leu
				(nmoles/g)		
Control		2.06±0.02	1	64	175	67
	0.5 mM	2.04±0.12	517	831	265	89
2-AB						
	1.0 mM	1.90±0.06	1042	1908	303	105
	2.0 mM	1.77±0.24	1978	3431	408	119
Imazaquin	0.08 µM	0.73±0.20	20	122	156	66
	0.4 µM	0.53±0.07	46	139	101	61
	2.0 µM	0.43±0.07	178	293	74	52
	10 µM	0.37±0.07	270	214	77	48

[a]Growth rate is a measure of the rate of leaf elongation between 8 and 24 hours after initiation of treatment. Treatments were applied continuously via the nutrient solution.

thase; (ii) leucine inhibits isopropylmalate synthase, the first enzyme unique to the biosynthesis of leucine; and (iii) ile inhibits threonine dehydratase, first enzyme unique to the biosynthesis of ile (Bryan, 1980). Levels of valine and leucine increased in the plants exposed to ile, presumably due to an increased flow of carbon through the pathway leading to the biosynthesis of these amino acids (Table 1). Pyruvate concentration is over 25-fold higher than the concentration of 2-KB (our unpublished data), yet a decline in 2-KB concentration causes increased flow of carbon to valine and leucine biosynthesis. This result suggests that plant AHAS has a higher affinity for 2-KB than pyruvate. A higher affinity for 2-KB than pyruvate has also been shown for AHAS II and AHAS III from *Escherichia coli* (Barak, Chipman & Gollop, 1987).

Valine-treated plants showed elevated levels of val and leu and reduced levels of ile (Table 1). The increased level of leucine in these plants is an expected result because ketoisovalerate produced from valine could be used in the pathway leading to the biosynthesis of leucine. Reduction in the pool of ile is probably due to the inhibition of AHAS activity by val and leu which reduces carbon flow to ile. This conclusion is supported by the accumulation of 2-KB and 2-AB in these plants.

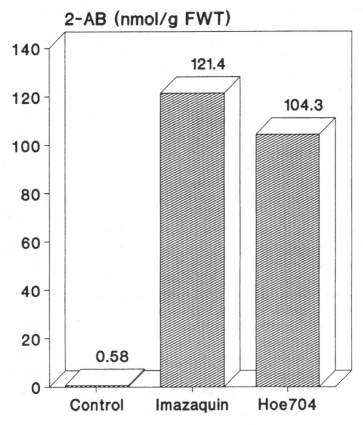

Fig. 4. Effects of imazaquin and Hoe704 on accumulation of 2-AB in hydroponically grown corn seedlings.

Pools of only val and leu but not of ile are reduced due to inhibition of AHAS by imazaquin (Table 1). This result suggests that imazaquin is a better inhibitor of the acetolactate producing- than the acetohydroxy-acid producing-reaction catalysed by AHAS. Therefore, carbon continues to flow in the pathway leading to ile biosynthesis. However, when val was fed alone or in combination with imazaquin, reduced pools of ile were observed. This effect was most dramatic when plants were treated with a combination of val and imazaquin. This treatment reduced the levels of ile by nearly 70% compared to the untreated plant. Such reduction in the pools of ile was not simply due to a competition from val and leu for the last aminotransferase reaction leading to ile biosynthesis because there was no accumulation of 2-oxo-3-methyl valerate, the keto acid leading to isoleucine (data not presented). Therefore, a combination of val, leu (synthesized *in vivo* from fed val) and imazaquin appears to completely inhibit AHAS thereby preventing carbon flow to ile. This conclusion is also supported by a greater accumulation of 2-KB and 2-AB owing to val+imazaquin treatment than with imazaquin treatment alone (Table 1, data for 2-KB not presented).

Ile appears to interfere with the flow of carbon from val to leu since the pools of leu were lower in all treatments containing ile (Table 1). This reduction in leu pool could be due to the inhibition of isopropylmalate synthase by ile. It is also possible that the same transaminase carries out the last reaction in the biosynthetic pathways of val, leu and ile. Therefore, a high concentration of ile would saturate the transaminase (in the back reaction) and prevent the transamination of 2-ketoisocaproate to leucine. This conclusion is supported by the accumulation of 2-ketoisocaproate in the treatments which included ile.

These findings also help explain why val and ile supplementation is not as good as supplementation with all three amino acids in reversing the growth inhibitory effects of imazaquin. Val treatment alone allows high accumulation of leu because val is converted to ketoisovalerate, an intermediate in leu biosynthesis. However, a combination of ile with val may prevent the conversion of val to leu. This effect is even more pronounced when ile and val are provided to the plant in combination with imazaquin. The leu pool in this treatment is lower than the pool present in control plants. Therefore, these plants do not grow at the same rate as the plants supplemented with all three amino acids.

Inhibition of threonine dehydratase is herbicidal

Wittenbach, Rayner & Schloss (1992) have suggested that TD is not a good herbicide target site based on the phenotype of isoleucine auxotrophs which lack TD. However, we wanted to explore whether or not inhibition of TD would be herbicidal. The only known inhibitor of plant TD is isoleucine. An antimetabolite, 2-(1-cyclohexen-3(R)-yl)-S-glycine (CHG), has been identified as an inhibitor of bacterial TD (Fig.5; Scannell & Preuss, 1974). We tested this compound and found that it is a competitive inhibitor of TD from Black Mexican Sweet corn cells. In the presence of 4 mM threonine (the Km for the corn enzyme), 0.4 mM CHG inhibited enzyme activity by 50%. CHG also inhibits TD *in vivo* because it can reduce the accumulation of 2-KB/2-AB in imazaquin treated plants. This result is possible only by inhibition of TD *in vivo*.

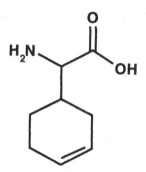

Fig. 5. Structure of 2-(1-cyclohexen-3(R)-yl)-S-glycine.

CHG was phytotoxic to *Arabidopsis* (Fig.6). This phytotoxicity was reversed when the growing media was supplemented either with isoleucine alone or with 2-AB. Leu, val or the other 17 protein amino acids supplementation had no effect on CHG-phytotoxicity. These results show that the phytotoxicity of CHG was due to its inhibition of TD and that this could be a potential herbicide target site.

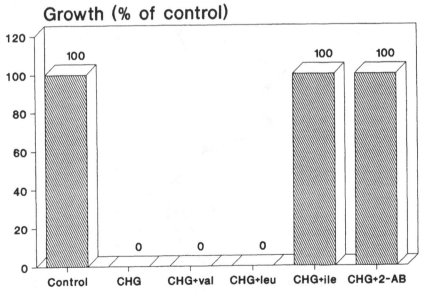

Fig. 6. Effects of 2-(1-cyclohexen-3(R)-yl)-S-glycine on growth of *Arabidopsis* in the presence of val, leu, ile or 2-AB. 2-(1-cyclohexen-3(R)-yl)-S-glycine and amino acids were used at 1 mM each.

Identification of an isoleucine feedback-insensitive form of threonine dehydratase

During our studies on the role of 2-KB and 2-AB in the herbicidal effects of AHAS-inhibiting herbicides, we decided to examine TD activity in various tissues of tomato. An isoleucine-sensitive form of threonine dehydratase activity was detected in flowers, fruits and leaves which shows that TD is important for isoleucine biosynthesis in all of these plant parts (Szamosi, Shaner & Singh, 1993). TD activity was high in the young leaves (Fig.7) and the activity declined with leaf age. However, there was a sudden rise in the specific activity of the enzyme in the older senescing leaves. This behaviour is quite different from AHAS in the same plant tissue. Further characterization of the TD activity revealed that this activity is a result of the appearance

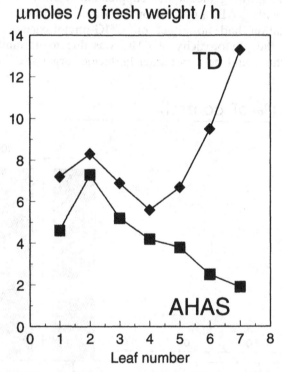

Fig. 7. Threonine dehydratase (◆) and acetohydroxyacid synthase (■) activity in the crude extracts of tomato leaves. Leaf numbers represent the leaves from the top (1; the youngest) to the bottom (7; the oldest) of a tomato plant.

of an isoleucine-insensitive form of the enzyme. Besides isoleucine insensitivity, the enzyme from old leaves differed from the enzyme from young leaves in pH optima, substrate saturation and molecular weight (Table 3). There is precedence for a biodegradative form of TD in microorganisms (Umbarger and Brown 1956, Whiteley & Tahara 1966), in parasitic and saprophytic plants (Kagan, Sinelnikova & Kretovich, 1969, Madan &, Nath 1983), and in peas (Tomova *et al.*, 1969). Our results suggest that the two forms of enzyme identified in tomato are separate gene products. One gene expresses the isoleucine-sensitive form of the enzyme that is responsible for isoleucine biosynthesis. The other gene expresses the isoleucine-insensitive form of the enzyme which may play a role in leaf senescence and is not involved in isoleucine biosynthesis but in degradation. These conclusions are supported by our observation that inhibition of AHAS by imazaquin causes accumulation of 2-KB/2-AB in young leaves only (Fig.8). Similarly, Hoe704 treatment causes accumulation of acetolactate only in young leaves (36.3 AU/g fresh wt in the young leaves vs 3.6 AU/g fresh weight in the old leaves). These results suggest that there is very little carbon flowing through the branched chain amino acid biosynthetic pathway in old tomato leaves.

Then why do these senescing tomato leaves contain such high levels of isoleucine-insensitive TD activity? Proteins break down during leaf senescence. The liberated amino acids are then catabolized to produce ammonia which is used to produce glutamine, the amino acid transported out of senescing leaves (Thimann, 1980). However, the mechanism of deamination of amino acids in the older leaves has not been elucidated. An isoleucine-insensitive form of TD present in senescing leaves could degrade threonine and serine to release ammonia which

Table 3. *Properties of threonine dehydratase from the young leaves (fresh wt = 31 mg/leaf; chlorophyll = 6.9 mg/g fresh weight) and the old leaves (fresh wt = 1450 mg/leaf; chlorophyll = 0.5 mg/g fresh weight) of tomato*

Property	Young leaves	Old leaves
pH optima	5.5–6.0; 9.5	9.0–9.5
K_m threonine (mM)	0.25	0.25
K_m serine (mM)	1.7	0.25
Inhibition by isoleucine (%)	>80	<10
Molecular weight (Da)	370 000	200 000

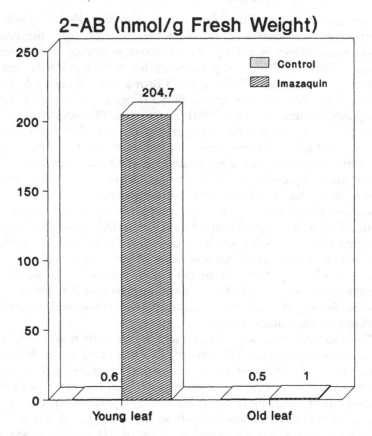

Fig. 8. Accumulation of 2-aminobutyrate in imazaquin-treated young leaves (fresh wt = 31 mg/leaf; chlorophyll = 6.9 mg/g fresh weight) and the old leaves (fresh wt = 1450 mg/leaf; chlorophyll = 0.5 mg/g fresh weight) of tomato.

can be used by glutamine synthetase to produce glutamine. It is believed that glutamine synthetase activity in the senescing leaf is sufficient to produce glutamine from all of the N released during protein hydrolysis (Storey & Beevers, 1978). Therefore, the biodegradative form of TD may metabolize both threonine and serine to release nitrogen for use in young, growing tissues. Similar forms of other enzymes for release of ammonia from other amino acids may be present in senescing leaves. Discovery of such enzymes will complete the missing link of how ammonia is released from amino acids before incorporation into glutamine in senescing leaves.

Summary

In conclusion, intensive research on the branched chain amino acids in plants over the last 10 years has greatly increased our understanding of this important pathway and how it is regulated. Questions still remain, however, on how this pathway is integrated with the other metabolic pathways, and on the factors which determine the potency of inhibitors of different enzymes within the pathway. Continued research on these questions should greatly increase our understanding of amino acid biosynthesis in plants in the future.

References

Barak, Z., Chipman, D. & Gollop, N. (1987). Physiological implications of the specificity of acetohydroxy acid synthase enzymes of enteric bacteria. *Journal of Bacteriology,* **169**, 3750–6.

Bryan, J. (1980). Synthesis of aspartate family and branched chain amino acids. In *The Biochemistry of Plants*, ed. B. Miflin, pp. 403–53. New York: Academic Press.

Kagan, Z. S., Sinelnikova, E.M. & Kretovich, W.L. (1969). L-threonine dehydratases of flowering parasitic and saprophytic plants. *Enzymologia, 36*, 335–52.

Langzagorta, A., de la Torre, J.M. & Aller, P. (1988). The effect of butyrate on cell cycle progression in *Allium cepa* root meristems. *Physiologia Plantarum, 72*, 775–81.

LaRossa, R.A. & Van Dyk, T.K. (1987). Metabolic mayhem caused by 2-ketoacid imbalances. *Bioessays, 7*, 125–30.

LaRossa, R.A., Van Dyk, T.K. & Smulski, D.R. (1987). Toxic accumulation of α-ketobutyrate caused by inhibition of the branched-chain amino acid biosynthetic enzyme acetolactate synthase in *Salmonella typhimurium*. *Journal of Bacteriology,* **169**, 1372–8.

LaRossa, R.A., Van Dyk, T.K. & Smulski, D.R. (1990). A need for metabolic insulation: lessons from sulfonylurea genetics. In *Biosynthesis of Branched Chain Amino Acids*, ed. Z. Barak, D.M. Chipman & J.V. Schloss, pp. 109–21. Weinheim, Germany: VCH Publishers.

Madan, V.K. & Nath, M. (1983). Threonine(serine) dehydratase from *Cuscuta campestris* Yunck. *Biochemica Physiologia Pflanzen,* **178**, 43–51.

74 B.K. SINGH, I. SZAMOSI AND D. SHANER

Reid, R.J., Field, L.D. & Pitman, M.S. (1985). Effects of external pH, fusicoccin and butyrate on the cytoplasmic pH in barley root tips measured by ^{31}P-nuclear magnetic resonance spectroscopy. *Planta,* **166,** 341–7.

Rhodes, D., Hogan, A., Deal, L., Jamieson, G. & Howarth, P. (1987). Amino acid metabolism of *Lemna minor* L. II. Responses to chlorsulfuron. *Plant Physiology,* **84,** 775–80.

Scannell, J. & Pruess, D. (1974). Naturally occurring amino acid and oligopeptide antimetabolites. In *Chemistry and Biochemistry of Amino Acids, Peptides and Proteins.* vol. 3, pp. 189–244.

Schloss, J.V. (1989). Origin of the herbicide binding site of acetolactate synthase. In *Prospects for Amino Acid Biosynthesis Inhibitors in Crop Protection and Pharmaceutical Chemistry,* ed. L.G. Copping, J. Dalziel & A.D. Dodge, pp 147–52. British Crop Protection Council Monograph No. 42.

Schloss, J.V. & Aulabaugh, A. (1989). Acetolactate synthase and ketol-acid reductoisomerase: A search for reason and a reason for search. In *Biosynthesis of Branched Chain Amino Acids,* ed. Z. Barak, D.M. Chipman & J.V. Schloss, pp. 329–56, Weinheim, Germany: VCH Publishers.

Schultz, A., Sponemann, P., Kocher, H. & Wengenmeayer, F. (1988). The herbicidally active experimental compound Hoe 704 is a potent inhibitor of the enzyme acetolactate reductoisomerase. *FEBS Letters,* **238,** 375–8.

Shaner, D. & Singh, B. (1992). How does inhibition of amino acid biosynthesis kill plants? In *Biosynthesis and Molecular Regulation of Amino Acids in Plants,* ed. B.Singh, H.Flores & J.Shannon, pp. 174–83. Rockville: American Society of Plant Physiologists.

Storey, R. & Beevers, L. (1978). Enzymology of glutamine metabolism related to senescence and seed development in the pea (*Pisum sativum* L.). *Plant Physiology,* **61,** 494–500.

Szamosi, I., Shaner, D. & Singh, B. (1993). Identification and characterization of a biodegradative form of threonine dehydratase in senescing tomato leaf. *Plant Physiology,* **101,** 999–1004.

Thimann, K.V. (1980). The senescence of leaves. In *Senescence in Plants,* ed. K.V. Thimann, pp. 85–116. Boca Raton: CRC Press.

Tomova, W.S., Kagan, Z.S., Blekhman, G.I. & Kretovich, W.L. (1969). Biodegradative L-threonine dehydratase from pea seedlings. *Biokhimiya,* **34,** 266–72.

Umbarger, H.E. & Brown, B. (1956). Threonine deamination in *Escherichia coli.* II. Evidence for two L-threonine deaminases. *Journal of Bacteriology,* **73,** 105–13.

Van Dyk, T.K. & LaRossa, R.A. (1986). Sensitivity of a *Salmonella typhimurium* AspC mutant to sulfometuron methyl, a potent inhibitor of acetolactate synthase II. *Journal of Bacteriology,* **165,** 386–92.

Van Dyk, T.K. & LaRossa, R.A (1987). Involvement of ACK–PTA operon products in α-ketobutyrate metabolism by *Salmonella typhimurium*. *Molecular and General Genetics,* **207**, 435–40.

Van Dyk, T.K., Smulski, D.R. & Chang, Y.Y. (1987). Pleiotropic effects of poxA regulatory mutations of *Escherichia coli* and *Salmonella typhimurium*, mutations conferring sulfometuron methyl and α-ketobutyrate hypersensitivity. *Journal of Bacteriology,* **169**, 4540–6.

Whiteley, H.R. & Tahara, M. (1966). Threonine deaminase of *Clostridium tetanomorphum*. I. Purification and properties. *Journal of Biological Chemistry,* **241**, 4881–9.

Wittenbach, V., Aulabaugh, A. & Schloss, J. (1991). Examples of extraneous site inhibitors and reaction intermediate analogues: acetolactate synthase and ketol-acid reductoisomerase. In *Pesticide Chemistry. Advances in International Research, Development and Legislation*, ed. H Frehse, pp. 151–60. Weinheim, New York: VCH publishers.

Wittenbach, V., Rayner, D. & Schloss, J. (1992). Pressure points in the biosynthetic pathway for branched chain amino acids. In *Biosynthesis and Molecular Regulation of Amino Acids in Plants*, ed. B. Singh, H. Flores & J. Shannon, pp. 69–88. Rockville: American Society of Plant Physiologists.

T. DAVID UGALDE, SHERIDAN E.
MAHER, NINO E. NARDELLA and
ROGER M. WALLSGROVE

Amino acid metabolism and protein deposition in the endosperm of wheat: synthesis of proline via ornithine

Background

Between 75% and 85% of the mature wheat grain is starch, so above all else, yield is a measure of the whole-plant processes that culminate in starch deposition in the grain. Protein percentage, on the other hand, is a ratio value, and whilst not independent of yield is obviously an expression of nitrogen metabolism. The rates and durations of both starch and protein deposition in the endosperm of wheat all appear to be independent events controlled by separate mechanisms (Jenner, Ugalde & Aspinall, 1991). It is this independence that gives the opportunity to manipulate specific responses in the plant that culminate in starch and protein deposition, whether the attempts at improvement be genetic or agronomic.

The relationship between substrate supply and dry matter deposition is different for starch and protein, and the responses change during grain development (Jenner *et al.*, 1991). During the grain filling stage (10–15 days after anthesis until the onset of maturity), the rate of starch deposition in healthy plants is mainly influenced by sink-limited factors, that is by factors that operate within or close to the grain itself. By contrast, deposition of protein is influenced to a much greater extent by source-limited factors, that is by factors of supply. Increasing amino acid supply to developing grains leads directly to increases in protein deposition. Within this context, however, the levels of substrate within the endosperm (sucrose and amino acids respectively) appear inconsistent with what may be expected. There are enough amino acids within wheat endosperm to sustain about $1\frac{1}{2}$ days of protein deposition, while in contrast there is only enough sucrose for a few hours of deposition of starch (Ugalde & Jenner, 1990*a,b*). Hence, as far as regulation of protein deposition is concerned, the important questions become why are there such high levels of amino acids in wheat endosperm, and why is there still a substrate-driven response?

One explanation may be that the rate of protein deposition is limited by one, or perhaps a few amino acids almost independent of the level of total amino acids in the pool. For each of the main amino acids, Ugalde and Jenner (1990*b*) calculated its rate of supply from external sources, its rate of incorporation into protein, and by difference, the rate of synthesis required in the endosperm. Levels of amino acids in the soluble pool within the endosperm were measured also, and turn-over times within this soluble pool were calculated. Most amino acids do not fit with expected characteristics of one that may limit the rate of the whole process, but a number do, and these divide clearly into four structurally related groups: proline, branched-chain amino acids (isoleucine, leucine, valine), aromatic amino acids (phenylalanine, tyrosine), and sulphur-containing (cysteine, methionine) (Ugalde, 1993). Further work (Ugalde, 1993) tested which group or groups of amino acids responded to substrate supply. As amino acid supply to developing wheat endosperm was increased, the rate of protein deposition increased concurrent with small shifts in the proportional content of some amino acids in the deposited protein. Glutamine and proline were the only ones that increased. Hence proline was the one amino acid with characteristics suggesting a limiting role that increased in the deposited protein when a limit to protein deposition was alleviated.

Requirement for proline in wheat endosperm

Very little, if any, proline is transferred from the vegetative tissue of wheat to the developing grain (Ugalde & Jenner, 1990*a*). Within the grain, amino acids move longitudinally in the vascular bundle at the base of the crease and radially within the nucellar projection and across the endosperm cavity (Ugalde & Jenner, 1990*b*). The cavity is a large and experimentally-accessible apoplast between the maternal and filial tissues of the grain. In extracts from both the main vascular bundle and the endosperm cavity, proline is less than 2% (mol mol^{-1}) of the total complement of amino acids.

Proline makes up about 12% (mol mol^{-1}) of the protein deposited in wheat endosperm, and so clearly there is the need for high synthetic capacity for proline within the endosperm itself. Glutamine makes up about 30% (mol mol^{-1}) of the amino acids entering wheat endosperm, and the amount of glutamate is small (3%; Ugalde & Jenner, 1990*a*). Within the endosperm, there is a flow of nitrogen away from glutamine to glutamate, presumably catalysed by glutamate synthase (EC 1.4.1.14; Sodek & da Silva, 1977), such that the concentration of glutamate is twice that of gluta-mine. Glutamate is the obvious substrate for synthesis of proline.

Evidence for proline synthesis in wheat endosperm via ornithine

There are two possible pathways for synthesis of proline from glutamate, via glutamyl phosphate or via ornithine (Fig. 1). We have assayed the activities of several of the enzymes potentially involved in extracts of peeled endosperms (harvested 22 days after anthesis). For all enzymes, crude homogenates were clarified by centrifugation and desalted on small columns of Sephadex G-25 prior to assay.

The glutamate acetyltransferase assay contained endosperm extract, L-acetylornithine or L-acetylglutamate, and L-glutamate or L-ornithine, plus 1 mM L-proline and/or 1 mM L-arginine in some reactions. Zero time, minus enzyme and minus amino acid controls were run. The ornithine aminotransferase assay contained endosperm extract, 2-oxoglutarate, and ornithine. L-Proline (2 mM) was included in some assays. Similar controls to the acetyltransferase assay were included. The 'pyrroline-5-carboxylate synthetase' assay contained endosperm extract, ATP, MgCl$_2$, NADPH, and glutamate. Similar controls were used.

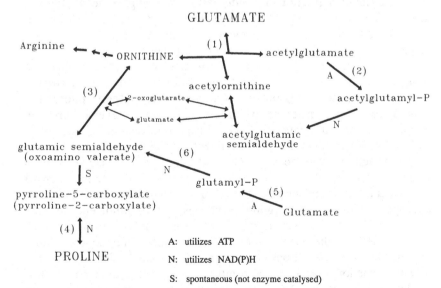

Fig. 1. Possible alternative pathways for synthesis of proline and arginine from glutamate in developing wheat endosperm: (1) Glutamate acetyltransferase; (2) Acetylglutamate kinase; (3) Ornithine aminotransferase; (4) Pyrroline-5 (-2)-carboxylate reductase (EC 1.5.1.2/ 1.5.1.1); (5) Glutamate kinase; (6) Glutamate semialdehyde dehydrogenase.

For all these assays, reaction mixtures were incubated at 30 °C and aliquots removed sequentially. These were added to equal volumes of cold ethanol, thoroughly mixed and then centrifuged. Amino acids in the resulting supernatants were analysed by reverse-phase HPLC of the PITC derivatives, using a gradient system based on the Waters procedure for analyses of hydrolysates (Bidlingmeyer, Cohen & Tarvin, 1984; Nardella, unpublished observations).

Acetylglutamate kinase was assayed by following the production of acetylglutamyl hydroxamate in the presence of excess hydroxylamine, essentially as described previously for the assay of aspartate kinase (EC 2.7.2.4; Relton *et al.*, 1988). L-proline (1 mM) was included in some assays, and zero time and minus enzyme controls were used. Glutamate kinase was assayed in the same way, substituting L-glutamate for L-acetylglutamate, and glutamine synthetase (EC 6.3.1.2) (transferase reaction) was assayed as previously described (Wallsgrove *et al.*, 1987).

For the separation of acetylglutamate kinase and glutamine synthetase, crude supernatant was fractionated with ammonium sulphate. In one experiment a fraction of crude extract was applied to a Mono Q anion exchange column (Pharmacia FPLC system) and chromatographed using the conditions previously established for the separation of glutamine synthetase isoenzymes (Wallsgrove *et al.*, 1987).

Several of the enzymes required for the synthesis of proline via ornithine were detected. In contrast, no evidence was found for enzyme activities associated with the direct pathway via glutamyl phosphate. Glutamate acetyltransferase when assayed in the reverse direction (appearance of glutamate) showed a linear rate over a 1 hour assay, and glutamate production was dependent on the presence of ornithine in the reaction. The activity, 68 pkat endosperm^{-1}, converts to a potential capacity of 5.9 μmol endosperm^{-1} day^{-1}, greatly in excess of the calculated requirement for proline synthesis of 0.17 μmol endosperm^{-1} day^{-1}, and arginine synthesis of 0.02 μmol endosperm^{-1} day^{-1} (Table 1). Neither arginine nor proline inhibited this activity when added to the reaction medium at 1 mM final concentration. Under the assay conditions used here (phosphate buffer pH7.2), the forward reaction (synthesis of ornithine) was less active; about 30% of the back reaction. Ornithine production was absolutely dependent on the presence of glutamate.

Incubation of cell-free endosperm extracts with glutamate in the presence of ATP and NADPH ('pyrroline-5-carboxylate synthetase') did not lead to any detectable synthesis of proline or pyrroline-5-carboxylate. The instability of the intermediates (glutamyl phosphate,

Table 1. *Predicted requirements of proline and arginine synthesis within developing wheat endosperm for protein deposition, and* in vitro *activities of enzymes in the alternative pathways*

	μmol endosperm^{-1} day^{-1}
Required synthesis of proline	0.17
Required synthesis of arginine	0.02
Glutamate acetyltransferase activity	5.9
Acetylglutamate kinase activity	1.6
Ornithine aminotransferase activity	2.3
'Pyrroline-5-carboxylate synthetase' activity	Not detected

Predicted requirements taken from Ugalde & Jenner 1990*a*.

glutamic semialdehyde) in neutral aqueous solutions could mean that they were formed but then broke down, or competing ATPases and diaphorases may have rapidly degraded the added co-factors, and so restricted or masked the activity of proline synthesis enzymes. This is perhaps unlikely, as there was no difficulty in detecting acetylglutamate kinase and glutamine synthetase, both of which require ATP. For whatever reason, we were unable to detect any proline synthesis via the direct (glutamyl phosphate) pathway from glutamate (Fig. 1). Nor could we detect any accumulation of ornithine in these reaction mixtures, as might be expected if glutamic semialdehyde had been synthesised and then transaminated (glutamate being present in excess).

The second step in the ornithine pathway, catalysed by acetylglutamate kinase (Fig. 1) was also active in developing endosperms (18.8 pkat endosperm^{-1}). This activity is distinct from glutamine synthetase and readily separated from it by ammonium sulphate fractionation. More than 90% of the glutamine synthetase activity precipitated at 40–60% saturation, whereas 70% of all acetylglutamate kinase activity precipitated out at 0–40% saturation. All the apparent 'glutamate kinase' activity found was associated with glutamine synthetase, and no part of this activity or the acetylglutamate kinase was inhibited by 1 mM proline. Fractionation of endosperm extract on an anion exchange column revealed only a single peak of 'glutamate kinase' activity, entirely coincident with the single peak of glutamine synthetase. Acetylglutamate kinase activity was not inhibited by methionine sulphoximine at concentrations up to 1mM, that is at concentrations known to totally inhibit glutamine synthetase (see for example Ericson, 1985).

Ornithine can be metabolized to either arginine or proline. For proline synthesis, an aminotransferase reaction is required, leading to either pyrroline-5-carboxylate or pyrroline-2-carboxylate. A 2-oxoglutarate-dependent ornithine aminotransferase was found in the endosperm extracts, with a pH optimum of about 8 as reported for the enzyme from other plant sources (Splittstoesser & Fowden, 1973; Lu & Mazelis, 1975). Essentially, no activity could be detected at pH 7. We did not attempt to distinguish between the products pyrroline-5-carboxylate and pyrroline-2-carboxylate, as either can serve as proline precursors (MacHolan, Zobac & Hekelova, 1965). Proline, up to 2 mM, did not inhibit this activity.

Ornithine aminotransferase activity (as measured by the ornithine + 2-oxoglutarate-dependent appearance of glutamate) was linear for at least 40 minutes under the conditions used, with an activity of 27 pkat endosperm^{-1}. This converts to a potential capacity of 2.33 μmol endosperm^{-1} day^{-1}, and as for glutamate acetyltransferase and acetylglutamate kinase (above) this is well in excess of the calculated requirement for the synthesis of proline. Table 1 summarizes the requirements for proline and arginine synthesis in wheat endosperm, and the *in vitro* activities of the assayed enzymes.

Implications for amino acid metabolism in the endosperm of wheat

Arginine makes up only about 3.5% (mol/mol) of the amino acids in protein deposited in wheat endosperm, and about 60% of the arginine required for protein deposition is delivered to the grain from the rest of the plant. The requirement for synthesis of arginine in the endosperm is thus relatively small.

Arginine is derived from glutamate via ornithine, and is an end-product not serving as a precursor for synthesis of any other amino acid. The activities of glutamate acetyltransferase and acetylglutamate kinase reported here imply abundant capacity for alternative use of ornithine or any other of the intermediates. Ornithine aminotransferase activity would produce glutamate semialdehyde or 2-oxo-5-amino valerate for subsequent cyclisation (probably non-enzymic) to pyrroline carboxylic acids and reduction to proline. By contrast, no activity was found in the pathway for proline synthesis via glutamyl phosphate in our cell-free extracts. This route certainly is active in mammalian and microbial systems, and has been assumed to be the major route in plants (Rayapati & Stewart, 1986; Steff & Vasakova, 1984). However, evidence presented to date for the activity of glutamate kinase in plants is largely circumstantial and unsatisfactory; one problem being inability to distinguish between glutamine synthetase and glutamate kinase. Glutamine synthetase is present in all plant tissues and is capable of catalysing a 'glutamate kinase' activity *in vitro* (when hydroxylamine is used to trap the glutamyl phosphate), although there is no indication that this enzyme operates as a kinase *in vivo*. Our failure to identify a direct route to proline in wheat endosperm may, of course, simply reflect an inherent difficulty in extracting one or more of the enzymes concerned, or an instability of one of the enzymes. The analogous aspartate kinase, for example, is very unstable *in vitro* and difficult to assay in crude plant extracts (Relton *et al.*, 1988). There is also the possibility that the synthesis of proline in different tissues proceeds via different pathways, depending on the tissue and the purpose for synthesis as, for instance, for the very large amounts of proline that may be synthesized in vegetative tissue as a response to stress. Genes coding for the enzymes of both potential pathways have been cloned from plant tissues, including a glutamate kinase gene (Garcia-Rios *et al.*, 1991), implying that in some tissues at least the direct route to proline operates. A bifunctional enzyme has also been reported in *Vigna aconitifolia*, and the gene cloned, which catalyses both the glutamate kinase and P-5-C synthetase activities (Hu, Delauney & Verma, 1992). It has also been suggested that the pathway depends on the level of N-availability, with the ornithine pathway operating when adequate N is available. High N is reported to enhance ornithine transaminase gene expression and depress P-5-C synthetase gene expression in mothbean seedlings (Delauney *et al.*, 1993); the developing cereal endosperm can be considered an N-rich tissue.

There is some evidence (Kueh *et al.*, 1984) that synthesis of proline in plants is regulated by feedback inhibition by proline, although the target enzyme has not been identified. In our studies, we were unable to demonstrate any significant effect of proline on any of the ornithine pathway enzymes, and so the nature of any metabolic regulation remains unclear. If proline synthesis is limiting in endosperm tissue, it may be that there is no requirement for subtle feedback regulation in the pathway in this tissue. The regulation (and indeed the pathway) of proline in vegetative tissues of wheat remains to be clarified.

From the results presented here, we propose that the developing wheat endosperm has the capacity to synthesize all of its proline requirement, in addition to arginine, via the ornithine pathway. While we cannot discount the presence and operation of the direct pathway from glutamate to proline, to date, we can find no evidence for the activity of the requisite enzymes. In leaf tissue, the enzymes of the ornithine pathway are all localized in the plastid (Taylor & Stewart, 1981), so the precise intracellular location of ornithine transaminase and the pyrroline carboxylate reductases may give more indication of the involvement of ornithine in proline synthesis. Co-location of acetylornithine transaminase and ornithine aminotransferase may allow coupled cycling of glutamate and 2-oxoglutarate as indicated in Fig. 1.

Conclusions

There appear to be abundant levels of amino acids for protein deposition in the endosperm of wheat, but a substrate-driven rate response is still observed. It appears that the rate of protein deposition is influenced, perhaps determined by the levels of certain amino acids that are in short supply, almost independent of the total amount in the soluble pool. Proline is the second most abundant amino acid in wheat-grain protein (12% mol mol^{-1}), yet very little, if any, proline is transferred from the vegetative tissues to the grain.

Some enzymes of possible biosynthetic pathways of proline synthesis in developing wheat endosperm have been assayed in cell-free extracts. No evidence was found for direct synthesis of pyrroline-5-carboxylate via glutamate kinase and pyrroline-5-carboxylate synthetase. In contrast, glutamate acetyltransferase (EC 2.3.1.35), acetylglutamate kinase (EC 2.7.2.8) and ornithine aminotransferase (EC 2.6.1.13) were found with activities greatly in excess of those required to account for the predicted rate of synthesis of proline and arginine for protein deposition. There was no evidence for inhibition of any of these enzymes by proline, and so the mechanism of any metabolic control of these pathways in the endosperm remains unknown. We propose that developing wheat endosperm has the capacity to synthesize all of the proline it requires via the ornithine biosynthetic pathway.

Acknowledgements

This work was supported by the Australian Grains Research and Development Corporation and the UK Agricultural and Food Research Council. We are grateful to David Banfield for assistance with growth room facilities.

References

Bidlingmeyer, B.A., Cohen, S.A. & Tarvin, T.L. (1984). Rapid analysis of amino acids using pre-column derivatisation. *Journal of Chromatography* **336**, 93–104.

Delauney, A.J., Hu, C.A., Kishor, P.B.K. & Verma, D.P.S. (1993). Cloning of ornithine δ-aminotransferase cDNA from *Vigna aconiti-*

folia by *trans*-complementation in *Escherichia coli* and regulation of proline biosynthesis. *Journal of Biological Chemistry,* **268,** 18673–8.

Ericson, M.C. (1985). Purification and properties of glutamine synthetase from spinach leaves. *Plant Physiology,* **79,** 923–7.

Garcia-Rios, M.G., LaRoss, P.C., Bressen, R.A., Csonka, L.N. & Hanquier, J.M. (1991). Cloning by complementation of the gamma-glutamyl kinase gene from a tomato expression library. *Abstracts of the Third International Plant Molecular Biology Congress,* Tucson, USA.

Hu, C.A., Delauney, A.J. & Verma, D.P.S. (1992). A bifunctional enzyme (Δ^1-pyrroline-5-carboxylate synthetase) catalyses the first two steps in proline biosynthesis in plants. *Proceedings of the National Academy of Sciences USA,* **89,** 9354–8.

Jenner, C.F., Ugalde, T.D. & Aspinall, D (1991). The physiology of starch and protein deposition in the endosperm of wheat. *Australian Journal of Plant Physiology,* **18,** 211–26.

Kueh, J.H., Hill, J.M., Smith, S.J. & Bright, S.W.J. (1984). Proline biosynthesis in a proline-accumulating barley mutant. *Phytochemistry,* **23,** 2207–10.

Lu, T. & Mazelis, M. (1975). L-Ornithine: 2-oxoacid aminotransferase from squash (*Curcubita pepo* L) cotyledons. *Plant Physiology,* **55,** 502–6.

MacHolan, L., Zobac, P. & Hekelova, P. (1965). Activity, utilization and metabolism of 5-amino-2-oxovaleric acid (γ-pyrroline-2-carboxylic acid) in pea seedlings and baker's yeast. *Hoppe Seylers Zeitschrift für Physiologie Chemie,* **349,** 97–106.

Rayapati, P.J. & Stewart, C. (1986). Gamma-glutamyl kinase activity from wilted barley leaves. *Plant Physiology,* **80** suppl., abstract 305.

Relton, J., Bonner, P., Wallsgrove, R.M. & Lea, P.J. (1988). Physical and kinetic properties of lysine-sensitive aspartate kinase purified from carrot cell suspension culture. *Biochimica et Biophysica Acta,* **953,** 48–60.

Sodek, L. & da Silva, W.J. (1977). Glutamate synthase: a possible role in nitrogen metabolism of the developing maize endosperm. *Plant Physiology,* **60,** 602–5.

Splittstoesser, W.E. & Fowden, L. (1973). Ornithine transaminase from *Cucurbita maxima* cotyledons. *Phytochemistry,* **12,** 785–90.

Steff, M., & Vasakova, L. (1984). Regulation of proline-inhibitable glutamate kinase of winter wheat leaves by monovalent cations and L-proline. *Collection of Czechoslovac Chemical Communications,* **49,** 2698–708.

Taylor, A.A. & Stewart, G.R. (1981). Tissue and subcellular localization of enzymes of arginine metabolism in *Pisum sativum. Biochemistry and Biophysiology Research Communications,* **101,** 1281–9.

Ugalde, T.D. (1993) A physiological basis for genetic improvement to nitrogen harvest index in wheat. In *Genetic Aspects of Plant*

Mineral Nutrition, eds P.J. Randall *et al.*, pp. 301–9. Kluwer Academic Publishers.

Ugalde, T.D. & Jenner, C.F. (1990*a*). Substrate gradients and regional patterns of dry matter deposition within developing wheat endosperm. II. Amino acids and protein. *Australian Journal of Plant Physiology*, **17**, 395–406.

Ugalde, T.D. & Jenner, C.F. (1990*b*). Route of substrate movement into wheat endosperm. II. Amino acids. *Australian Journal of Plant Physiology*, **17**, 705–14.

Wallsgrove, R.M., Turner, J.C., Hall, N.P., Kendall, A.C. & Bright, S.W.J. (1987). Barley mutants lacking chloroplast glutamine synthetase – biochemical and genetic analysis. *Plant Physiology*, **83**, 155–8.

STEPHEN RAWSTHORNE,
ROLAND DOUCE and DAVID OLIVER

The glycine decarboxylase complex in higher plant mitochondria: structure, function and biogenesis

Glycine decarboxylation and photorespiratory metabolism in C_3 plants

In higher plants which carry out C_3 photosynthesis, photosynthetic and photorespiratory metabolism is based on the action of ribulose-1,5-bisphosphate (RuBP) carboxylase/oxygenase and the regeneration of its substrate ribulose-1,5-bisphosphate by the reductive pentose phosphate pathway (RPP) or Calvin cycle. These reactions occur in the chloroplast. The product of CO_2 fixation by RuBP carboxylase is two molecules of 3-phosphoglycerate (a three-carbon compound; hence C_3 photosynthesis) which is either exported from the chloroplast as triose phosphate for sucrose synthesis in the cytosol or metabolized to form starch within the chloroplast or used for regeneration of RuBP. Oxygen competes with CO_2 for the active site of RuBP carboxylase/ oxygenase leading to an oxygenase reaction which reduces the rate of CO_2 assimilation. The products of the oxygenase reaction are 3-phosphoglycerate and phosphoglycolate (a two-carbon compound). The production of phosphoglycolate represents a drain of carbon away from the RPP and to recover this carbon the phosphoglycolate is metabolized through a series of reactions involving enzymes in the chloroplasts, peroxisomes, and mitochondria. In the course of this pathway two molecules of glycine (i.e. four carbon atoms) are metabolized to one molecule of serine, CO_2 and NH_3. Serine is metabolized further to 3-phosphoglycerate and so three out of four carbon atoms entering the pathway are returned to the RPP (Husic, Husic & Tolbert, 1987; Ogren, 1984). The light energy requirement of photosynthesis and photorespiration is for the synthesis of ATP and NADPH via the electron transport pathway of the chloroplast. These compounds are then used to drive reactions of the RPP.

It is the light-dependent loss of CO_2 from a leaf which leads to the term photorespiration. When the CO_2 available to RuBP carboxylase/

oxygenase becomes progressively limiting in the light (e.g. due to environmental stresses which lead to limitation of CO_2 diffusion through the stomates) the light energy captured by the leaf is dissipated through the operation of the RPP and the oxygenase reaction. Photorespiratory metabolism is therefore able to prevent the formation of the excited triplet state of chlorophyll and excess reactive O_2 species (superoxide radicals and singlet oxygen) which would be damaging to the chloroplast. Because of its major role in photorespiratory metabolism, this review will address our current understanding of glycine oxidation by mitochondria in photosynthetic tissues and how this process is mediated by metabolic interaction and gene expression.

Subunit structure and reaction mechanism

The first reactions in the metabolism of photorespiratory glycine are carried out by glycine decarboxylase (GDC) which is also described as the glycine cleavage system. This multienzyme complex is located in the mitochondrial matrix and it catalyses the oxidative decarboxylation and deamination of glycine with the formation of CO_2, NH_3 and N^5N^{10}-methylene-5,6,7,8-tetrahydropteroyl-glutamate ($H_4PteGlu_n$ where n = number of glutamate residues) (Neuburger, Bourguignon & Douce, 1986). The methylene group is transferred from the latter molecule to a second molecule of glycine to form serine and $H_4PteGlu_n$ in a reaction catalysed by serine hydroxymethyltransferase (SHMT). Considering the high rate of glycine oxidation in leaf mitochondria and the relatively low content of folate compounds in mitochondria (see below) it is clear that the availability of $H_4PteGlu_n$ for GDC and its recycling through the SHMT reaction may be a critical step for glycine oxidation during photorespiration.

Glycine decarboxylase has been purified from plant mitochondria (Walker & Oliver, 1986a; Bourguignon, Neuburger & Douce, 1988), animals (Kikuchi, 1973; Kikuchi & Hiraga, 1982) and bacteria (Klein & Sagers, 1966; Kochi & Kikuchi, 1969) and consists of four protein components which have been named P-protein (a homodimer containing pyridoxal phosphate), H-protein (a monomeric lipoamide-containing protein), T-protein (a monomer catalysing the $H_4PteGlu_n$-dependent step of the reaction), and L-protein (dihydrolipoamide dehydrogenase; a homodimer containing flavin adenine dinucleotide [FAD]). The subunit molecular weights of these proteins are given in Table 1. Lipoic acid is attached to the H-protein via an amide linkage to the ε-amino group of a lysine residue (lysine-63 on the pea H-protein; Mérand et al., 1993). The H-protein has been crystallized and the X-ray crystal

Table 1. *Physical characterization of the proteins of the glycine decarboxylase complex and their respective cDNAs*

Characteristic	Component protein			
	P-protein	L-protein	T-protein	H-protein
Precursor protein				
Number of amino acids	1057	501	408	165
Presequence				
Number of amino acids	86	31	30	34
Mature protein				
MW[a] by SDS-PAGE[b] (kDa)	98	59	45	15
MW by sequence/MS[c] (kDa)	105	50	41	14
Number of amino acids	971	470	378	131
Gene copies per haploid genome	2	1–2	1	1
mRNA size (kbp)	3.4	1.8	1.4	0.7

[a] Molecular weight; [b] Sodium dodecylsulphate-polyacrylamide gel electrophoresis; [c] Mass spectrometry.

structure has been determined at 2.6 Å by multiple isomorphous replacement techniques. The lipoate cofactor attached to lys[63] is located in the loop of a hairpin configuration. This means that the lipoate group is not buried and has freedom to interact readily with the other three proteins in the multienzyme complex (Pares *et al.*, 1994).

Glycine decarboxylase from plant leaf mitochondria is closely related to similar enzyme complexes found in bacteria such as *Peptococcus glycinophilus* (Klein & Sagers, 1966) and *Athrobacter globiformis* (Kochi & Kikuchi, 1969) and in the mitochondria of animal tissues (Kikuchi, 1973). The dihydrolipoamide dehydrogenase component of GDC, which contains, in addition to FAD, a redox active cystine residue, is also a component (called E3) of a family of multienzyme complexes composed of pyruvate, 2-oxoglutarate and branched-chain 2-oxoacid dehydrogenases (Reed, 1974; Bourguignon *et al.*, 1992; Turner, Ireland & Rawsthorne, 1992*b*). Serine hydroxymethyltransferase has been purified from plant leaf mitochondria and is a 220 kDa homotetramer with a subunit molecular weight of 53 kDa (Bourguignon *et al.*, 1988). The enzyme purified from pea leaf mitochondria is unrelated immunologically to the enzyme from root tissues and a second, and also unrelated, form of the enzyme occurs in leaves (Turner *et al.*, 1992*c*; and see Hiltz & Ireland in this volume). Serine hydroxymethyltransferase also requires pyridoxal-phosphate and each subunit has a single pyridoxal-phosphate bound as a Schiff base to an ε-amino group of a lysyl residue (Schirch, 1984).

The GDC reaction begins with the amino group of glycine forming a Schiff base with the pyridoxal-phosphate of the P-protein. The carboxyl group of glycine is lost as CO_2 and the remaining methylamine moiety is passed to the lipoamide cofactor of the H-protein. The lipoamide-bound methylamine group is shuttled to the T-protein where the methylene carbon is transferred to $H_4PteGlu_n$ to produce $CH_2H_4PteGlu_n$ and the amino nitrogen is released as NH_3. Only the 6S stereoisomer of $CH_2H_4PteGlu_n$, presumably the natural occurring form, is the substrate for this reaction in vitro (Besson et al., 1993). The last step of the reaction is the oxidation of the resulting dihydrolipoamide of the H-protein by the L-protein with the sequential reduction of FAD and NAD^+. During the early steps of H-protein isolation, the P- and H-proteins react together in the presence of glycine. Under these conditions H-protein with the methylamine intermediate bound to its lipoamide accumulates in the medium at the expense of the oxidized H-protein (Neuburger, Jourdain & Douce, 1991). The methylamine intermediate, which is a rather stable structure, can then be separated easily from the oxidized H-protein on ion-exchange chromatography (Neuburger et al., 1991). All the reactions catalysed by the glycine cleavage system are fully reversible. For example, the H- and L-proteins together catalyse the reversible exchange of electrons between NADH and lipoamide bound to the H-protein (Neuburger et al., 1991).

The equilibrium constant of the leaf mitochondrial SHMT suggests that during photorespiration, the reaction must be permanently pushed toward the formation of serine (the unfavourable direction) to allow the recycling of $H_4PteGlu_n$ necessary for the operation of the T-protein component of GDC. Serine hydroxymethyltransferase and the T-protein of the glycine cleavage system display higher affinities for $H_4PteGlu_n$ substrates containing more than three glutamic residues (Besson et al., 1993). In this context, analysis of the folates present in pea leaf mitochondria reveals a pool of polyglutamates dominated by tetra and pentaglutamates (Besson et al., 1993). The importance of polyglutamate derivatives in the folate metabolism of mitochondria is consistent with earlier observations that mutations affecting the generation of conjugated folates resulted in auxotrophies for methionine and products of one-carbon metabolism such as thymidine and purine (Cossins, 1987). The polyglutamate chain of plant mitochondrial folates could play an important role in the coordination of related enzyme activities such as those catalysed by the T- and serine hydroxymethyltransferase proteins.

Indeed, folylpolyglutamates are known to increase the efficiency of sequential folate-dependent enzymes by enhancing the 'channelling' of intermediates between the active sites of protein complexes (Schirch & Strong, 1989). In other words, folate polyglutamate intermediates do not equilibrate with the bulk solvent but are directly transferred between the active sites of the different proteins, a situation which facilitates the establishment of the steady state. In support of this hypothesis, Besson *et al.* (1993) have shown that the tetrameric SHMT and the monomeric T-proteins (quantification of the T- and SHMT proteins indicates that they represent respectively 10 and 2% of the soluble matrix proteins; Bourguignon *et al.*, 1988) are potentially able to bind twice the actual folate content of the matrix space (on a per mg of protein basis the total folate content in the matrix space is approximately 0.80 nmol mg^{-1}). This indicates that the polyglutamates of leaf mitochondria are probably bound largely to the active sites of these two predominant folate-dependent enzymes. Being bound rather than free may afford better protection to these very labile and oxidizable tetrahydrofolate compounds. There is some evidence that the negatively charged α-carboxyl groups of the poly-γ-glutamate chain could bind at specific points, such as basic groups of the proteins (Usha, Savithri & Rao, 1992). In this context, it is notable that regions of the amino acid sequence of the pea T-protein are characterized by the alignment of basic residues (Bourguignon *et al.*, 1993).

The rate of glycine release during the course of photorespiration is as much as 50% of the photosynthetic rate of about 3 μmol CO_2 fixed mg^{-1} chlorophyll min^{-1} and some 10 times the rate of normal tricarboxylic acid cycle activity. To accomplish rates of glycine oxidation which are rapid enough to cope with all the glycine molecules flooding out of the peroxisomes the glycine cleavage system linked to SHMT is present at tremendously high concentrations. In fact, it comprises about half of the soluble proteins in the matrix of mitochondria from fully expanded green leaves (Oliver *et al.*, 1990). This is in contrast with the situation observed in mammalian mitochondria where GDC represents a minute fraction of the total matrix protein (Douce & Neuburger, 1989). The high protein concentration (0.2 g GDC per ml of matrix volume) in plant mitochondria leads to the formation of a loose multienzyme complex with approximate subunit ratios of 2 P-protein dimers : 27 H-protein monomers : 9 T-protein monomers : 1 L-protein dimer (Oliver *et al.*, 1990; Douce *et al.*, 1991). The enzymological properties of this loose complex are very different from those of the dissociated form of GDC which occurs at low protein concentrations (Oliver *et al.*, 1990). In the dissociated state the H-protein acts as a mobile co-substrate that commutes between the other three enzymes and shows typical Michaelis–Menten substrate kinetics. In the loose complex the H-protein no longer acts as a substrate but is an integral part of the enzyme (Oliver et al., 1990).

Metabolic control of glycine decarboxylation

There is no evidence that the plant GDC activity in mature leaves is affected by light, by reversible covalent modification, by control proteins or by proteolytic activation. Serine is known to inhibit the glycine cleavage reaction competitively with respect to glycine (K_i serine = 4 mM; K_m glycine = 6 mM) and it appears to bind to the P-protein (Oliver & Sarojini, 1987). More importantly though, the catalytic activity of GDC is strongly regulated by the NADH/NAD$^+$ molar ratio which affects the L-protein directly and hence regeneration of the oxidized lipoyl moiety which is bound to the H-protein. Neuburger et al. (1986) have shown that NADH is a competitive inhibitor with respect to NAD$^+$ (K_m NAD$^+$ = 75 μM; K_i NADH = 15 μM). This means that increasing the ratio of NADH to NAD$^+$ in the matrix space will result in a logarithmic increase in inhibition of enzyme activity. NADH generated in the matrix space during the course of glycine oxidation must therefore be reoxidized if the photorespiratory cycle is to continue. There are several possible mechanisms for the oxidation of NADH produced by glycine oxidation and these cannot be considered to be mutually exclusive. In the first, NADH produced could be reoxidized very rapidly by oxaloacetate (OAA), owing to the tremendous excess of malate dehydrogenase located in the matrix space working in the reverse direction. A very powerful phthalonate-sensitive OAA carrier (K_m OAA = 5 μM; V_{max}=700 nmol mg^{-1} protein min^{-1}) has been characterized in all the plant mitochondria isolated so far (Douce, 1985; Ebbighausen, Chen & Heldt 1985). This OAA carrier would facilitate a rapid malate-OAA transport shuttle, the equivalent of which is not found in mammalian mitochondria. Operation of this shuttle would enable the transfer of reducing equivalents generated in the mitochondria during glycine oxidation to the peroxisomal compartment for the reduction of β-hydroxypyruvate (Krömer, Hanning & Heldt, 1992 & references therein). Indeed, in the presence of oxaloacetate, the glycine cleavage reaction in isolated mitochondria is able to operate faster than when the NADH is reoxidized exclusively via the mitochondrial electron transport chain operating under state 3 conditions, i.e. a non-limiting ADP concentration and the coupling of electron transport to ATP production (Lilley, Ebbighausen & Heldt,

1987). This second possible fate of NADH has been demonstrated by the photorespiration-dependent production of ATP in isolated leaf protoplasts (Gardeström & Wigge, 1988). A consideration to be taken into account regarding *in vivo* glycine oxidation via the respiratory chain, is that this process is dependent upon the ATP being recycled back to ADP at a rate sufficient to account for the potential rate of glycine-dependent oxidative phosphorylation. However, it is known that the rate of respiratory electron transport is limited by the low concentration of ADP which is available for oxidative phosphorylation (Hooks *et al.*, 1989; Dry & Wiskich, 1982, Wiskich & Meidan, 1992). If such a limitation were to occur under photorespiratory conditions, it would lead to an increase in the matrix NADH/NAD$^+$ ratio and thus reduce the rate of glycine cleavage.

A limitation to the rate of electron transport because of low ADP availability could of course, be negated if glycine oxidation was uncoupled from oxidative phosphorylation. Two possibilities exist which would enable this to occur. First, the oxidation of NADH could be linked to the non-phosphorylating alternative pathway. However, the results of Douce, Moore and Neuburger (1977) indicate that for green leaves there would be insufficient alternative pathway activity to completely support *in vivo* rates of photorespiratory glycine oxidation. Second, Wiskich and Meidan (1992) have proposed that the rate of glycine oxidation can be very rapid even under state 4 (i.e. ADP limited) conditions because of complex interactions between the electron transport chain, the proton motive force and the permeability of the mitochondrial membrane to protons. The net result of these interactions is that proton movement across the mitochondrial membrane becomes uncoupled from ATP production while allowing the rapid transfer of electrons from NADH to O_2 (Wiskich & Meidan, 1992)

Evidence to date has shown that glycine oxidation inhibits concurrent oxidation of other mitochondrial substrates but is itself not affected by these substrates (Day & Wiskich, 1981; Day, Neuburger & Douce 1985). The preferential oxidation of glycine could be achieved by a dominance of complex I over both complex II and the external NADH dehydrogenase of the respiratory chain and the ability of GDC to compete favourably at the level of matrix NAD$^+$ due to its low K_m relative to other mitochondrial dehydrogenases (Neuburger *et al.*, 1986; Bourguignon *et al.*, 1988). However, the situation may not be as simple as this and interactions between the oxidation of glycine and TCA cycle intermediates by isolated mitochondria are complex (Wiskich & Meidan, 1992). To account for this complexity, it has been proposed that the NADH produced by GDC has access to electron transport chains which are in some way spatially separated within a mitochondrion from those which are accessible to NADH produced from activity of the TCA cycle (Wiskich & Meidan, 1992). Clearly, resolving the precise mechanisms involved in the metabolism of NADH which is produced by GDC remains an important question although determining its fate *in vivo* will be very difficult. Nevertheless, in green leaves the rapid utilization of NADH and the immediate utilization of NH_3 (via glutamate synthase and glutamine synthetase operating in a concerted manner) and CO_2 (via RuBP carboxylase) during the course of glycine oxidation continuously shift the equilibrium of the GDC reaction towards serine formation even though the reactions are readily reversible *in vitro*.

Finally, compounds that react either with the lipoamide cofactor of the H-protein like arsenite, or with the pyridoxal phosphate of the P-protein like carboxymethoxylamine, methoxylamine and acethydrazide, strongly inhibit the glycine cleavage system (Sarojini & Oliver, 1985). Such compounds acting at the level of the protein components of GDC would therefore exhibit herbicidal potency. Indeed, for mutants of *Arabidopsis thaliana* in which GDC activity was deficient, photosynthesis was unimpaired in non-photorespiratory conditions but was irreversibly inhibited in atmospheres that allowed the rapid production of glycolate by chloroplasts (Somerville & Ogren, 1982; Somerville & Somerville, 1983).

Organ and cellular distribution of glycine decarboxylase in C₃ plants

Most of the studies on the distribution of GDC between and within plant organs, and the glycine oxidation capacities of mitochondria isolated from these organs have been made using C_3 plants. In these plants, the capacity for glycine oxidation is several fold higher in mitochondria isolated from photosynthetic organs than in those isolated from non-photosynthetic tissues, e.g. roots and leaf veins (Gardeström, Bergman & Ericson, 1980; Walton & Woolhouse, 1986), where activity is virtually undetectable. Whilst this distribution of glycine decarboxylation at the organ level is to be expected because of the role of GDC in photorespiration, it should be remembered that organs are in general highly differentiated and contain many different cell types. In the leaves of wheat, for example, only 50% of the cells are chloroplast-containing and therefore capable of photosynthesis (Jellings & Leech, 1982). The remaining cells are predominantly vascular, and in young expanding leaves, are still living (Jellings & Leech, 1982) and must contain mitochondria. It is clear that there are both spatial and temporal (i.e. developmental) influences on the presence of GDC in particular cell types within the leaf. In support of this, the work of Tobin and colleagues (Tobin *et al.*, 1989; Rogers *et al.*, 1991; Tobin & Rogers, 1992) has shown that GDC is detectable by immunogold labelling in all leaf cells but that the density of labelling is much greater in cells which contain chloroplasts. Even within chloroplast-containing cells there appears to be variation for the density of immunogold labelling for the presence of GDC (e.g. the immunolabelling density for the P-protein on mitochondria of stomatal guard cells in pea leaves is half that on mesophyll cells) (Tobin *et al.*, 1989). Tobin *et al.* (1989) have speculated that this variation might represent a relationship between the capacity of a cell for photosynthesis and therefore the requirement for photorespiratory capacity. More recent data suggest that there is a positive correlation between the activity of RuBP carboxylase and the expression of GDC genes in different tissue types (Oliver, unpublished; see below). Furthermore, the amount of GDC in mesophyll cell mitochondria also increases as a leaf develops as does the total amount of the enzyme in each photosynthetic cell (Tobin & Rogers, 1992 and see below).

Glycine decarboxylase in C_3–C_4 and C_4 plants

Not all plant species have C_3 photosynthesis and considerable attention has been paid to the cell-specific distribution and role of GDC in leaves of species which have C_3–C_4 intermediate or C_4 photosynthesis (for recent reviews see Edwards & Ku, 1987; Rawsthorne, 1992; Rawsthorne, Leegood & von Caemmerer, 1992). The gas exchange phenotype of C_3–C_4 intermediate photosynthesis is very characteristic. The CO_2 compensation point (Γ) of these intermediate species is significantly less than that of related C_3 species (typically 9–30 versus 45–55 μl CO_2 l^{-1} air, respectively) and it also decreases in response to increasing light intensity. In contrast, Γ of C_3 species is unaffected by light intensity. The low Γ of C_3–C_4 intermediate species is due to a light-dependent recapture mechanism which increases the efficiency of recapture of photorespired CO_2 before it leaves the leaf (Hunt, Smith & Woolhouse, 1987; Holbrook, Jordan & Chollet, 1985). The biochemical basis of this mechanism is the loss of GDC activity from the photosynthetic mesophyll cells of leaves of these species (Hylton *et al.*, 1988; Rawsthorne *et al.*, 1988a,b). It has been proposed that this results in the confinement of all release of photorespired CO_2 to the bundle sheath cells which surround the vascular tissue (Rawsthorne *et al.*,

1988a). These cells contain numerous mitochondria, all of which contain GDC, and which are in close association with chloroplasts. This distribution of the GDC within the leaf and the special anatomy of the bundle-sheath cells provides the potential for considerable enhancement of photorespiratory CO_2 recapture compared to that in a C_3 species.

The C_3–C_4 intermediate mechanism is known to occur in several species representing a range of genera from both monocotyledonous and dicotyledonous plants (Rawsthorne, 1992). However, the way in which glycine decarboxylase activity is lost from the mesophyll differs in different genera. Morgan, Turner & Rawsthorne (1993) have shown that the P-, L-, T- and H-proteins are all undetectable or present only at very low amounts in the mitochondria of mesophyll cells of C_3–C_4 intermediate species in the genera *Panicum* and *Flaveria*. In contrast, only the P-protein is absent from the mesophyll cell mitochondria of the C_3–C_4 species *Moricandia arvensis*. Furthermore, the study of *P. laxum* has revealed that there is yet further variation in the loss of GDC from the mesophyll of C_3–C_4 intermediates. This species has been classified previously as a C_3 plant. However, its leaf anatomy tends towards that of a C_3–C_4 intermediate with an increase in the number of organelles present in the bundle sheath cells compared to that in a related C_3 species (Brown *et al.*, 1983). Furthermore, the GDC immunolabelling density on the mitochondria in the mesophyll of *P. laxum* is only 50–60% of that on those in the bundle-sheath (Hylton *et al.*, 1988). These 'partial' C_3–C_4 characters and a reduced, light responsive Γ (Rawsthorne, unpublished observations) suggest that this species should be more closely grouped with related C_3–C_4 intermediate rather than C_3 species.

The complete loss of all the GDC subunits from the mesophyll in a given C_3–C_4 species is linked with the presence, in the same genus, of species with C_4 photosynthesis (Morgan *et al.*, 1993). In C_4 species, the GDC proteins and activity are also confined to the bundle-sheath cells (Hylton *et al.*, 1988; Ohnishi & Kanai, 1983; Petit & Cantrel, 1986; Morgan *et al.*, 1993). It has been proposed that C_3–C_4 intermediate species represent evolutionary intermediates in the transition between the older C_3 photosynthetic mechanism and the younger C_4 one (Edwards & Ku, 1987; Rawsthorne, 1992). By drawing upon the spectrum of C_3–C_4 phenotypes which occur across the different genera a sequence of discrete steps is apparent, each representing a progressive change towards C_4 photosynthesis (Fig. 1). Whether such a sequence actually occurred during the evolution of C_4 photosynthesis, and what the adaptive advantages of each step might have been, remain to be proven.

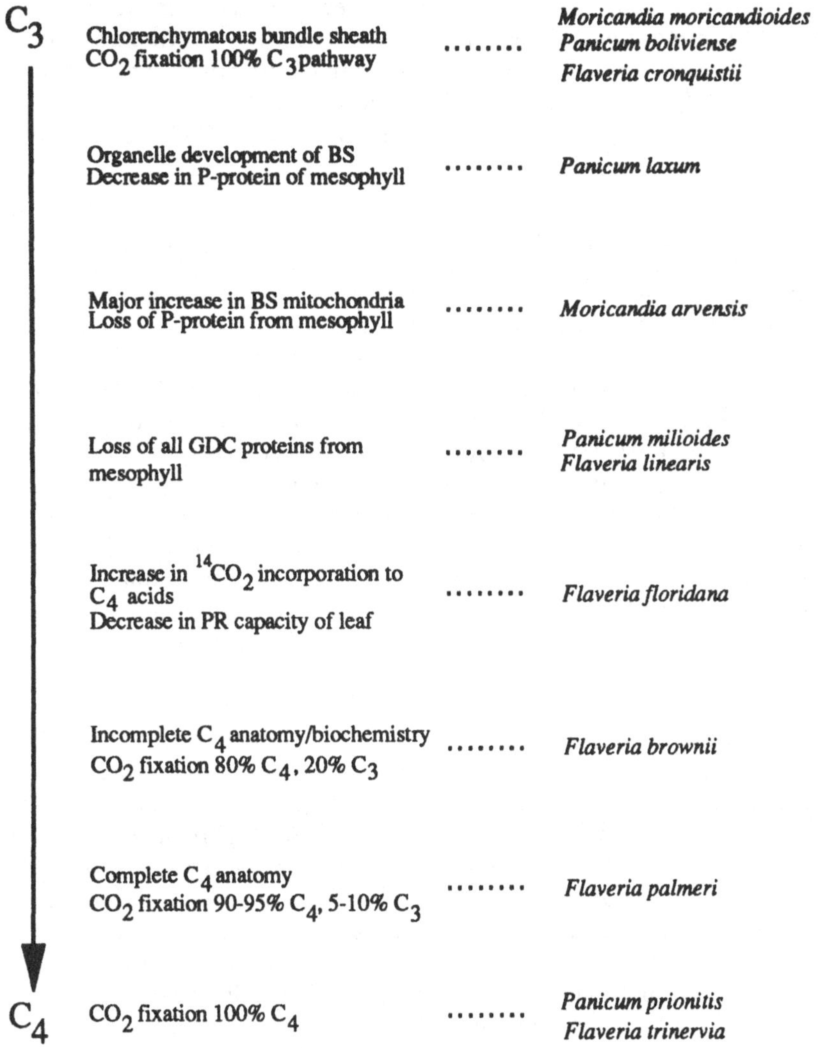

C_3

Chlorenchymatous bundle sheath
CO_2 fixation 100% C_3 pathway .... *Moricandia moricandioides*
Panicum boliviense
Flaveria cronquistii

Organelle development of BS
Decrease in P-protein of mesophyll .... *Panicum laxum*

Major increase in BS mitochondria
Loss of P-protein from mesophyll .... *Moricandia arvensis*

Loss of all GDC proteins from
mesophyll .... *Panicum milioides*
Flaveria linearis

Increase in $^{14}CO_2$ incorporation to
C_4 acids
Decrease in PR capacity of leaf .... *Flaveria floridana*

Incomplete C_4 anatomy/biochemistry .... *Flaveria brownii*
CO_2 fixation 80% C_4, 20% C_3

Complete C_4 anatomy
CO_2 fixation 90-95% C_4, 5-10% C_3 .... *Flaveria palmeri*

C_4

CO_2 fixation 100% C_4 .... *Panicum prionitis*
Flaveria trinervia

Fig. 1. A proposed series of evolutionary steps between the C_3 and C_4 photosynthetic mechanisms and the position which present day C_3–C_4 intermediate species might represent on this progression.

The distribution of serine hydroxymethyltransferase between bundle-sheath and mesophyll cells has also been studied in both C_3–C_4 and C_4 plants. In C_4 species the SHMT activity is much higher in the bundle-sheath than in the mesophyll cells (Ohnishi & Kanai, 1983). In the mesophyll of the C_3–C_4 intermediate species *M. arvensis* the amount of SHMT protein and its activity are reduced by between three- and five-fold relative to that in the bundle-sheath (Rawsthorne *et al.*, 1988*b*; Morgan *et al.*, 1993). Ku *et al.* (1992) have also shown that there is a progressive decrease in the extractable activity of SHMT from leaves of C_3, C_3–C_4, and C_4 species, respectively.

The genes encoding the glycine cleavage complex

cDNA clones have been isolated for the P-protein (Kim, Shah & Oliver, 1991; Turner, Ireland & Rawsthorne, 1992*a*), H-protein (Kim & Oliver, 1990; Macherel *et al.* 1990), T-protein (Bourguignon *et al.*, 1993, Turner *et al.*, 1993), L-protein (Turner *et al.*, 1992*b*; Bourguignon *et al.*, 1992), and SHMT (Turner *et al.*, 1992*c*) from peas and the H-protein from *Arabidopsis* (Srinivasan & Oliver, 1992). All of these were isolated by means of monospecific antibodies. In addition, genomic clones for the P-protein gene from *Moricandia* (Turner, unpublished observations), and the H-protein genes from pea (Macherel, Bourguignon & Douce, 1992), *Arabidopsis* (Srinivasan and Oliver, unpublished observations) and *Moricandia* (Franza & Rawsthorne, unpublished observations) have all been isolated. Major characteristics of the four genes which encode GDC, their mRNAs and the proteins encoded by them are summarized in Table 1.

All of these nuclear-encoded proteins are synthesized on 80S cytosolic ribosomes and imported into the mitochondria. Each of the GDC proteins is synthesized with an N-terminal leader sequence ranging from 30 amino acids for the T-protein to 86 amino acids for the P-protein. While these presequences lack a conserved amino acid sequence, they can all fold into an amphiphilic α-helix with the positive residues located along one face. Following uptake into the mitochondrial matrix, the presequence is removed by a specific signal peptidase and the mature proteins assemble into the active multienzyme complex. The pre-H-protein was synthesized by *in vitro* transcription and translation of the pea H-protein cDNA and used in uptake experiments with isolated pea leaf mitochondria. The precursor form of the protein was taken into the mitochondria and processed to the size of the mature protein (Srinivasan, Kraus & Oliver, 1992). Although equivalent work has not been done with the plant protein, the H-protein from bovine is lipoylated in the mitochondria following uptake and processing (Fujiwara, Okamura-Ikeda & Motokawa, 1992).

Glycine decarboxylase mutants have been identified for both *Arabidopsis thaliana* (Somerville & Ogren 1982) and barley (Blackwell,

Murray & Lea, 1990). While neither group of mutants has been characterized at the molecular level, the barley mutants show decreased levels
of both P-protein and H-protein. The number of copies per haploid
genome of the gene encoding the different GDC proteins differs, both
for each protein and also between separate reports for the same protein.
Macherel *et al.* (1990) and Kim & Oliver (1990) have suggested that
gdcH is a single copy gene and Bourguignon *et al.* (1992, 1993) have
proposed the same for *gdcL* and *gdcT*. Turner *et al.* (1992*a,b*) have
suggested that there are two copies of *gdcP* and *gdcL* (only one of
the *gdcL*s being expressed) while there are 2–3 copies of the gene
encoding the mitochondrial isoform of SHMT (Turner *et al.*, 1992*c*).
Notwithstanding these differences in copy numbers, all of the members
of each family map to a single locus in the pea genome suggesting
tight linkage of the members of each gene family (Turner *et al.*, 1993).
Sequence analysis of ten independent P-protein cDNA clones from
peas show a common 3' untranslated sequence providing some support
that either a single gene exists or is predominant in that plant.

Light-dependent expression of glycine decarboxylase

Green leaves have 5 to 10 times more GDC activity than etiolated
tissues (Tobin *et al.*, 1989; Walker & Oliver, 1986*b*). When etiolated
peas are transferred to light, there is a lag of approximately 4 hours
before the amount of GDC activity begins to increase. This increase
results from an elevation in the level of three of the four component
enzymes of the complex, the P- , H- , and T-proteins (Walker &
Oliver 1986*b*; Turner *et al.*, 1993). The L-protein is present at higher
initial concentrations in etiolated tissue and its level, therefore,
undergoes a smaller proportional increase (Turner *et al.*, 1992*b*; Bourguignon *et al.*, 1992). In wheat leaves the P- , H- , and T-proteins of
GDC accumulate along the developmental gradient from the young
tissue at the leaf base to older tissue at the tip (Rogers *et al.*, 1991).
Increases in the L-protein do occur in these developing wheat leaves
although the initial level of L-protein is much higher than that of the
others and the increase is far less marked (Rogers *et al.*, 1991). There
is some evidence that phytochrome is involved in the light-dependent
expression of GDC activity (Morohashi, 1987).

The light-dependent increases in the levels of the GDC proteins are
the result of increases in the amounts of the mRNAs encoding these
proteins. The mRNA levels for the P- , H- , and T-proteins are low
in etiolated peas. Following illumination of these plants, there is a 3
to 4 hour lag before the levels of these mRNAs increase (Fig. 1). In

fully green leaves, the level of these mRNAs is about 5 to 20 times greater than in etiolated tissues (Kim & Oliver, 1990; Kim et al., 1991; Turner et al., 1992a, 1993; Macherel et al., 1990). Careful comparisons of the abundance of these mRNAs during the greening process indicates that the P-protein mRNA may begin to accumulate sooner than the mRNAs for the other two genes (Turner et al., 1993). The mRNA for the L-protein does not change as dramatically reflecting the requirement for this protein as a component of other mitochondrial complexes in the dark (Turner et al., 1992b, Bourguignon et al., 1992).

Nuclei isolated from green pea leaves make about eight times more H-protein and P-protein mRNA than do nuclei isolated from etiolated plants (Srinivasan et al., 1992). In these run-on experiments, the nuclei will continue RNA synthesis for genes which are being transcribed at the time the nuclei were isolated, but will not initiate any new transcription of genes. Thus, run-on transcription measurements provide a quantitative snap shot of the amount of gene transcription when the nuclei were harvested. The observation that the increase in mRNA abundance is similar to the increase in transcription rate suggests that the light-dependent expression of gdcH and gdcP are controlled at the transcriptional level.

The lag between the onset of light and the increase in transcription is observed with several photosynthetic genes. It is usually assumed to result from the requirement for some as yet unidentified signal from fully matured chloroplasts before transcription of these light-induced genes is activated (Taylor, 1989; Susek & Chory, 1992). Support for this idea comes from two sources. When green pea plants are placed in the dark, the level of H-protein and P-protein mRNA drops dramatically to levels that are 10 to 20% those in etiolated tissues. After these plants are returned to light, the mRNA levels for both proteins rapidly return to the fully green level and the lag is reduced to less than one hour (Fig. 2), apparently because of the present of mature chloroplasts (Srinivasan, Berndt & Oliver, 1993). The second line of support comes from studies with the herbicide, Norflurazon. This chemical blocks carotenoid biosynthesis and, therefore, causes photobleaching and chloroplast destruction. Norflurazon-treated pea plants grown in bright light show many phytochrome-dependent changes including shorter stature and expanded leaves, but have a lower steady level state of H-protein mRNA in their leaves than do etiolated tissues (Srinivasan et al., 1993). The lack of mature chloroplasts appears to prevent the expression of the GDC genes.

The promoter region of the gdcH gene from Arabidopsis has been fused to the β-glucuronidase reporter gene (GUS) and this construct

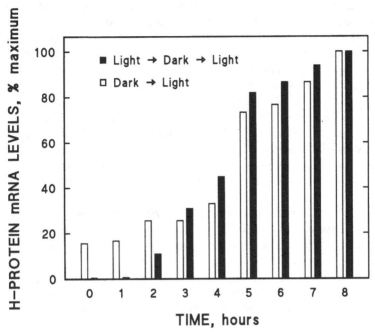

Fig. 2. Time course for the light-dependent increase in H-protein mRNA following illumination of etiolated pea plants (open bars) or reillumination of green plants that had been in the dark for three days (solid bars).

was transformed into tobacco (Srinivasan & Oliver, unpublished observations). When these transgenic tobacco plants were grown in the light, the GUS activity of the leaves was seven-fold greater than that from the leaves of etiolated plants. Plants transformed with constructs where the 5′ region of the promoter had been sequentially deleted had less GUS activity than plants carrying the full length *gdcH* promoter: GUS construct. Analysis of these deletions has revealed important *cis*-acting elements that have been implicated previously in the regulation of transcription of other light-induced genes (Gilmartin *et al.*, 1990).

Organ-specific expression of glycine decarboxylase

As discussed above, GDC activity is much higher in mitochondria isolated from leaves than in those isolated from petioles, stems, and roots. Analysis of the steady-state levels of mRNA and protein for

SHMT and the P- , T- and H-proteins in photosynthetic versus non-photosynthetic organs follows the same general pattern (Kim & Oliver, 1990; Macherel *et al.*, 1990; Bourguignon *et al.*, 1993; Turner *et al.*, 1992*a,c*). Comparable abundance of the mRNA encoding the L-protein in all organs provides further evidence for the role of the same L-protein in other mitochondrial, multisubunit dehydrogenase enzymes (Bourguignon *et al.*, 1992; Turner *et al.*, 1992*b*). Analysis of transgenic tobacco plants which contain the *gdcH* promoter:GUS fusion gene (Srinivasan & Oliver, unpublished observations) showed that the expression of GUS (as judged by the *in vitro* enzyme activity) in leaf lamina and veins, petioles, stems, roots, and seeds was approximately proportional to the RuBP carboxylase activity in each tissue type (Table

Table 2. *Enzyme activities of RuBP carboxylase and*
β-glucuronidase (GUS) in tissues of tobacco plants which were
transformed via Agrobacterium *with a* gdcH *promoter/GUS*
gene construct

Plant Tissue	RuBP carboxylase (nmol mg^{-1} protein min^{-1})	β-glucuronidase (nmol mg^{-1} protein h^{-1})
Leaf lamina	134	354
Leaf vein	70	163
Petiole	46	174
Stem	12	28
Root	1	15
Seed	5	2

2). Preliminary analysis of the promoter of the *gdcH* gene has identified specific regions that are responsible for the tissue specificity of expression (Srinivasan & Oliver, unpublished observations).

Summary

The study of the glycine decarboxylase complex of higher plant mitochondria has led to a considerable amount of information on the complexity of this multisubunit enzyme, its regulation by metabolism and by gene expression, and on the role it plays in photorespiration in C_3 and C_3–C_4 intermediate photosynthesis. However, much more work is required to answer the important questions of how the subunits of GDC interact within the complex and how the expression of the genes encoding the complex is regulated and coordinated at the organ-

and cell-specific levels. The abundance of GDC in plant mitochondria, the availability of specific antibodies and cDNA and genomic clones encoding the proteins of the enzyme, and the ability to manipulate the genes themselves and their expression *in planta* suggest that the answers to such questions can be realized.

Acknowledgements

The authors are grateful for the assistance of their numerous colleagues, without whom much of the work described above would not have been possible.

References

Besson, V., Rebeillé, F., Neuburger, M., Douce, R. & Cossins, E.A. (1993). Effects of tetrahydrofolate polyglutamates on the kinetic parameters of serine hydroxymethyltransferase from pea leaf mitochondria. *Biochemical Journal*, **292**, 425–30.

Blackwell, D., Murray, A.J.S. & Lea, P.J. (1990). Photorespiratory mutants of the mitochondrial conversion of glycine to serine. *Plant Physiology*, **94**, 1316–22.

Bourguignon, J., Macherel, D., Neuburger, M. & Douce, R. (1992). Isolation, characterization, and sequence analysis of a cDNA clone encoding L-protein, the dihydrolipoamide dehydrogenase component of the glycine cleavage system from pea-leaf mitochondria. *European Journal of Biochemistry*, **204**, 865–73.

Bourguignon, J., Neuburger, M. & Douce, R. (1988). Resolution and characterization of the glycine cleavage reaction in pea leaf mitochondria. Properties of the forward reaction catalyzed by glycine decarboxylase and serine hydroxymethyl-transferase. *Biochemical Journal*, **255**, 169–78.

Bourguignon, J., Vauclare, P., Mérand, V., Forest, E., Neuburger, M. & Douce, R. (1993). Glycine decarboxylase complex from higher plants: molecular cloning, tissue distribution and mass spectrometry analyses of the T protein. *European Journal of Biochemistry*, **217**, 377–86.

Brown, R.H., Bouton, J.H., Rigsby, L. & Rigler, M. (1983). Photosynthesis of grass species differing in carbon dioxide fixation pathways. VII. Ultrastructural characteristics of *Panicum* species in the *Laxa* group. *Plant Physiology* **71**, 425–31.

Cossins, E.A. (1987). Folate biochemistry and the metabolism of one-carbon units. In *The Biochemistry of Plants*, Vol. 11, ed. D.D. Davies, pp. 317–53. New York: Academic Press.

Day, D.A., Neuburger, M. & Douce, R. (1985). Interactions between glycine decarboxylase, the tricarboxylic acid cycle and the respiratory chain in pea leaf mitochondria. *Australian Journal of Plant Physiology*, **12**, 119–30.

Day, D.A. & Wiskich, J.T. (1981). Glycine metabolism and oxaloacetate transport by pea leaf mitochondria. *Plant Physiology*, **68**, 425–9.

Douce, R. (1985). *Mitochondria in Higher Plants: Structure, Function and Biogenesis.* New York: Academic Press.

Douce, R., Bourguignon, J., Besson, V., Macherel, D. & Neuburger, M. (1991). Structure and function of the glycine cleavage complex in green leaves. In *Molecular Approaches to Compartmentation and Metabolite Regulation*, ed. H.C. Huang & L. Taiz, pp. 59–74. The American Society of Plant Physiologists.

Douce, R., Moore, A.L. & Neuburger, M. (1977). Isolation and oxidative properties of intact mitochondria isolated from spinach leaves. *Plant Physiology*, **60**, 625–8.

Douce, R. & Neuburger, M. (1989). The uniqueness of plant mitochondria. *Annual Reviews of Plant Physiology and Plant Molecular Biology*, **40**, 371–419.

Dry, I.B. & Wiskich, J.T. (1982). The role of the external adenosine triphosphate/adenosine diphosphate ratio in the control of plant mitochondrial respiration. *Archives of Biochemistry and Biophysics*, **217**, 72–9.

Ebbighausen, H., Chen, J. & Heldt, H.W. (1985). Oxaloacetate translocator in plant mitochondria. *Biochemica et Biophysica Acta*, **810**, 184–99.

Edwards, G.E. & Ku, M.S.B. (1987). Biochemistry of C_3-C_4 intermediates. In *The Biochemistry of Plants*, Vol. 10, ed. M.D. Hatch M.D. & N.K. Boardman, pp. 275–325. London: Academic Press.

Fujiwara, K., Okamura-Ikeda, K. & Motokawa, Y. (1992). Expression of mature bovine H-protein of the glycine cleavage system in *Escherichia coli* and *in vitro* lipoylation of the apoform. *Journal of Biological Chemistry*, **267**, 20011–6.

Gardeström, P., Bergman, A. & Ericson, I. (1980). Oxidation of glycine via the respiratory chain in mitochondria prepared from different parts of spinach. *Plant Physiology*, **65**, 389–91.

Gardeström, P. & Wigge, B. (1988). Influence of photorespiration on ATP/ADP in the chloroplasts, mitochondria, and cytosol, studied by rapid fractionation of barley (*Hordeum vulgare*) protoplasts. *Plant Physiology*, **88**, 69–76.

Gilmartin, P.M., Sarokin, L., Memelink, J. & Chua, N-H. (1990). Molecular light switches for plant genes. *The Plant Cell*, **2**, 369–78.

Hooks, M.A., Clark, R.A., Neiman, R.H. & Roberts, J.K.M. (1989). Compartmentation of nucleotides in corn root tips studied by NMR and HPLC. *Plant Physiology*, **89**, 963–9.

Holbrook, G.P., Jordan, D.B. & Chollet, R. (1985). Reduced apparent photorespiration in the C_3-C_4 intermediate species, *Moricandia arvensis* and *Panicum milioides*. *Plant Physiology*, **77**, 578–83.

Hunt, S., Smith, A.M. & Woolhouse, H.W. (1987). Evidence for a light-dependent system for reassimilation of photorespiratory CO_2, which does not include a C_4 cycle, in the C_3–C_4 intermediate species *Moricandia arvensis*. *Planta*, **171**, 227–34.

Husic, D.W., Husic, H.D. & Tolbert, N.E. (1987). The oxidative photosynthetic carbon cycle or C_2 cycle. *CRC Critical Reviews in Plant Science*, **5**, 45–100.

Hylton, C.M. Rawsthorne, S., Smith, A.M., Jones, D.A. & Woolhouse, H.W. (1988). Glycine decarboxylase is confined to the bundle-sheath cells of C_3–C_4 intermediate species. *Planta*, **173**, 298–308.

Jellings, A.J. & Leech, R.M. (1982). The importance of quantitative anatomy on the interpretation of whole leaf biochemistry in species of *Triticum, Hordeum* and *Avena*. *New Phytologist*, **92**, 39–48.

Kikuchi, G. (1973). The glycine cleavage system: composition, reaction, mechanism and physiological significance. *Molecular and Cellular Biochemistry*, **1**, 169–87.

Kikuchi, G. & Hiraga, K. (1982). The mitochondrial glycine cleavage system. Unique features of the glycine decarboxylation. *Molecular and Cellular Biochemistry*, **45**, 137–49.

Kim, Y. & Oliver, D.J. (1990). Molecular cloning, transcriptional characterization, and sequencing of the cDNA encoding the H-protein of the mitochondrial glycine decarboxylase complex in peas. *Journal of Biological Chemistry*, **265**, 848–53.

Kim, Y., Shah, K. & Oliver, D.J. (1991). Cloning and light-dependent expression of the gene coding for the P-protein of the glycine decarboxylase complex from peas. *Physiologia Plantarum*, **81**, 501–6.

Klein, S.M. & Sagers, R.D. (1966). Glycine metabolism: 1. Properties of the system catalyzing the exchange of bicarbonate with the carboxyl group of glycine in *Peptococcus glycinophilus*. *Journal of Biological Chemistry*, **241**, 197–295.

Kochi, H. & Kikuchi, G. (1969). Reactions of glycine synthesis and glycine cleavage catalyzed by extracts of *Arthrobacter globiformis* grown in glycine. *Archives of Biochemistry and Biophysics*, **132**, 359–69.

Krömer, S., Hanning, I. & Heldt, H. (1992). On the sources of redox equivalents for mitochondrial oxidative phosphorylation in the light. In *Molecular, Biochemical and Physiological Aspects of Plant Respiration*, ed. H. Lambers & L.H.W. van der Plas, pp. 167–75. The Hague: SPB Academic Publishing.

Ku, M.S.B., Wu, J., Dai, Z., Scott, R.A., Chu, C. & Edwards, G.E. (1992). Photosynthetic and photorespiratory characteristics of *Flaveria* species. *Plant Physiology*, **96**, 518–28.

Lilley, R.M.C., Ebbighausen, H. & Heldt, H.W. (1987). The simultaneous determination of carbon dioxide release and oxygen update in suspensions of plant leaf mitochondria oxidizing glycine. *Plant Physiology*, **83**, 349–53.

Macherel, D., Bourguignon, J. & Douce, R. (1992). Cloning of the gene (gdc H) encoding H-protein, a component of the glycine decarboxylase complex of pea (*Pisum sativum*). *Biochemical Journal*, **286**, 627–30.

Macherel, D., Lebrun, M., Gagnon, J., Neuburger, M. & Douce, R. (1990). Primary structure and expression of H-protein, a component of the glycine cleavage system of pea leaf mitochondria. *Biochemical Journal*, **268**, 783–9.

Mérand, V., Forest, E., Gagnon, J., Monnet, C., Thibault, P., Neuburger, M. & Douce, R. (1993). Characterization of the primary structure of H-protein from *Pisum sativum* and location of a lipoic acid residue by combined LC-MS and LC-MS-MS. *Biological Mass Spectrometry*, **22**, 447–56.

Morgan, C.L., Turner, S.R. & Rawsthorne, S. (1993). Coordination of the cell-specific distribution of the four subunits of glycine decarboxylase and serine hydroxymethyltransferase in leaves of C_3–C_4 intermediate species from different genera. *Planta*, **190**, 468–73.

Morohashi, Y. (1987). Phytochrome-mediated development of glycine oxidation by mitochondria in cucumber cotyledons. *Physiologia Plantarum*, **70**, 46–50.

Neuburger, M., Bourguignon, J. & Douce, R. (1986). Isolation of a large complex from the matrix of pea leaf mitochondria involved in the rapid transformation of glycine into serine. *FEBS Letters*, **207**, 18–22.

Neuburger, M., Jourdain, A. & Douce, R. (1991). Isolation of H-protein loaded with methylamine as a transient species in glycine decarboxylase reactions. *Biochemical Journal*, **278**, 765–9.

Ogren, W. (1984). Photorespiration: Pathways, regulation, and modification. *Annual Reviews of Plant Physiology*, **35**, 415–42.

Ohnishi, J. & Kanai, R. (1983). Differentiation of photorespiratory activity between mesophyll and bundle sheath cells of C_4 plants. 1. Glycine oxidation by mitochondria. *Plant and Cell Physiology*, **24**, 1411–20.

Oliver, D.J., Neuburger, M., Bourguignon, J. & Douce, R. (1990). Interaction between the component enzymes of the glycine decarboxylase multienzyme complex. *Plant Physiology*, **94**, 833–9.

Oliver, D.J. & Sarojini, G. (1987). Regulation of glycine decarboxylase by serine. In *Progress in Photosynthesis Research*, Vol. III, ed. J. Biggins, pp. 573–6. The Hague: Martinus Nijhoff.

Pares, S., Cohen-Addad, C., Sieker, L., Neuburger, M. & Douce, R. (1994). X-ray structure determination at 2.6 Å-resolution of a lipoate containing protein: the H-protein of the glycine decarboxylase from pea leaves. *Proceedings of the National Academy of Sciences, USA,* (in press).

Petit, P. & Cantrel, C. (1986). Mitochondria from *Zea mays* leaf tissues: differentiation of carbon assimilation and photorespiratory activity between mesophyll and bundle sheath cells. *Physiologia Plantarum*, **67**, 442–6.

Rawsthorne, S. (1992). C_3–C_4 intermediate photosynthesis: linking gene expression to physiology. *Plant Journal*, **2**, 267–74.

Rawsthorne, S., Hylton, C.M., Smith, A.M. & Woolhouse, H.W. (1988*a*). Photorespiratory metabolism and immunogold localisation of photorespiratory enzymes in leaves of C_3 and C_3–C_4 intermediate species of *Moricandia*. *Planta*, **173**, 298–308

Rawsthorne, S., Hylton, C.M., Smith, A.M. & Woolhouse, H.W. (1988b) Distribution of photorespiratory enzymes between bundle-sheath and mesophyll cells in leaves of the C_3–C_4 intermediate species *Moricandia arvensis* (L.) DC. *Planta*, **176**, 527–32.

Rawsthorne, S., Leegood, R.C. & von Caemmerer, S. (1992). Metabolic interactions in leaves of C_3–C_4 intermediate plants. In *Plant Organelles: Compartmentation of Metabolism in Photosynthetic Tissue*, SEB Seminar Series Vol. 50, ed. A.K. Tobin, pp. 113–39. Cambridge: Cambridge University Press.

Reed, L.J. (1974). Multienzyme complexes. *Accounts in Chemical Research*, **7**, 40–6.

Rogers, W.J., Jordan, B.R., Rawsthorne, S. & Tobin, A.K. (1991). Changes in the stoichiometry of glycine decarboxylase subunits during wheat and pea leaf development. *Plant Physiology*, **96**, 952–6.

Sarojini, G. & Oliver, D.J. (1985). Inhibition of glycine oxidation by carboxyl-methoxylamine, methoxylamine and acethydroxide. *Plant Physiology*, **77**, 786–9.

Schirch, V. (1984). Folate in serine and glycine metabolism. In *Folate and Pterins, Vol. 1: Chemistry and Biochemistry of Folates*, ed. R.L. Blakley & S. Benkovic, pp. 399–431. New York: Wiley.

Schirch, V. & Strong, W.B. (1989). Interaction of folylpolyglutamates with enzymes in one-carbon metabolism. *Archives of Biochemistry and Biophysics*, **269**, 371–80.

Somerville, C.R. & Ogren, W.L. (1982). Mutants of the cruciferous plant *Arabidopsis thaliana* lacking glycine decarboxylase activity. *Biochemical Journal*, **202**, 373–80.

Somerville, S.C. & Somerville, C.R. (1983). Effects of oxygen and carbon dioxide on photorespiratory flux determined from glycine accumulation in a mutant of *Arabidopsis thaliana*. *Journal of Experimental Botany*, **34**, 415–24.

Srinivasan, R., Kraus, C. & Oliver, D.J. (1992). Developmental expression of the glycine decarboxylase multienzyme complex in greening pea leaves. In *Molecular, Biochemical, and Physiological Aspects of Plant Respiration*, ed. H. Lambers & L.H.W. van der Plas, pp. 323–34. The Hague: SPB Academic Publishing.

Srinivasan, R., Berndt, W.A. & Oliver, D.J. (1993). Coordinated expression of photosynthetic and photorespiratory genes. In *Plant*

108   S. RAWSTHORNE, R. DOUCE AND D. OLIVER

Mitochondria, ed. A. Brennicke & U. Kuck, pp. 241–50. Weinheim: VCH.

Srinivasan, R. & Oliver, D.J. (1992). H-protein of the glycine decarboxylase multienzyme complex, complementary DNA encoding the protein from *Arabidopsis thaliana*. *Plant Physiology*, **98**, 1518–19.

Susek, R.E. & Chory, J. (1992). A tale of two genomes: role of a chloroplast signal in coordinating nuclear and plastid genome expression. *Australian Journal of Plant Physiology*, **19**, 387–99.

Taylor, W.C. (1989). Regulatory interactions between nuclear and plastid genomes. *Annual Reviews of Plant Physiology and Plant Molecular Biology*, **40**, 211–33.

Tobin, A.K., Thorpe, J.R., Hylton, C.M. & Rawsthorne, S. (1989). Spatial and temporal influences on the cell-specific distribution of glycine decarboxylase in leaves of wheat and peas. *Plant Physiology*, **91**, 1219–25.

Tobin, A.K. & Rogers, W.J. (1992). Metabolic interactions of organelles during leaf development. In *Plant Organelles: Compartmentation of Metabolism in Photosynthetic Tissue*, SEB Seminar Series Vol. 50., ed. A.K. Tobin, pp. 293–323. Cambridge: Cambridge University Press.

Turner, S.R., Ireland, R.J. & Rawsthorne, S. (1992*a*). Cloning and characterization of the P subunit of glycine decarboxylase from pea. *Journal of Biological Chemistry*, **267**, 5355–60.

Turner, S.R., Ireland, R.J. & Rawsthorne, S. (1992*b*). Purification and primary amino acid sequence of the L subunit of glycine decarboxylase. *Journal of Biological Chemistry*, **267**, 7745–50.

Turner, S.R., Ireland, R.J., Morgan, C.L. & Rawsthorne, S. (1992*c*). Identification and localization of multiple forms of serine hydroxymethyltransferase in pea (*Pisum sativum*) and characterization of a cDNA encoding a mitochondrial isoform. *Journal of Biological Chemistry*, **267**, 13528–34.

Turner, S.R., Hellens, R., Ireland, R.J., Ellis, N. & Rawsthorne, S. (1993). The organisation and expression of the gene encoding the mitochondrial glycine decarboxylase complex and serine hydroxymethyltransferase in pea (*Pisum sativum*). *Molecular and General Genetics*, **236**, 402–8.

Usha, R., Savithri, H.S. & Rao, N.A. (1992). Arginine residues involved in binding of tetrahydrofolate to sheep liver serine hydroxymethyltransferase. *Journal of Biological Chemistry*, **267**, 9289–93.

Walker, J.L. & Oliver, D.J. (1986*a*). Glycine decarboxylase multienzyme complex. Purification and partial characterization from pea leaf mitochondria. *Journal of Biological Chemistry*, **261**, 2214–21.

Walker, J.L. & Oliver, D.J. (1986*b*). Light-induced increases in the glycine decarboxylase multienzyme complex from pea leaf mitochondria. *Archives of Biochemistry and Biophysics*, **248**, 626–38.

Walton, N.J. & Woolhouse, H.W. (1986). Enzymes of serine and glycine metabolism in leaves and non-photosynthetic tissues of *Pisum sativum* L.. *Planta*, **167**, 119–28.

Wiskich, J.T. & Meidan, E. (1992). Metabolic interactions between organelles in photosynthetic tissue: a mitochondrial overview. In *Plant Organelles: Compartmentation of Metabolism in Photosynthetic Tissue*, SEB Seminar Series Vol. 50, ed. A.K. Tobin, pp. 1–19. Cambridge: Cambridge University Press.

ROBERT J. IRELAND and DAVID A. HILTZ

Glycine and serine synthesis in non-photosynthetic tissues

The previous chapter has described the roles that glycine decarboxylase (GDC) and serine hydroxymethyltransferase (SHMT) play in converting photorespiratory-derived glycine to serine, primarily to recover some of the carbon from the glycolate produced by the oxidative reactions of RUBISCO in C3 plants. As well as being needed for protein synthesis, glycine and serine are precursors of a variety of molecules essential to the growth and development of plant tissues. Glycine is required for the synthesis of, amongst other things, glutathione, porphyrins such as leghemoglobin in nitrogen-fixing root nodules, and purines which are needed for nucleic acid synthesis and, in some nodules, for ureide production. Serine plays a similar role as a precursor of biomolecules, including phospholipids, tryptophan and cysteine, and, under some stress conditions is involved in the synthesis of glycine-betaine (see chapter by Gorham, this volume, and Rhodes & Hanson, 1993). The interconversion of glycine and serine, with the concomitant production of methylene tetrahydrofolate, is also important to the plant as a source of one carbon units (see Cossins, 1980). In C3 leaves, photorespiration provides substantial amounts of glycine and serine which can be used for such syntheses. In non-photosynthetic tissue such as roots and developing and germinating seeds, there is a need for glycine and serine just as in the leaves, but there is no photorespiration to produce high levels of glycine and serine. Alternative, non-photorespiratory routes for the production of glycine and serine exist in such tissues, as well as in the leaves of C3 plants. In the latter case, these pathways are probably of minor significance compared with the photorespiratory route during the daylight hours, but may be of more importance in the dark or early in leaf development.

This chapter will present a brief overview of what is known about glycine and serine synthesis in roots and seeds and compare this with the situation in leaves; the reader is referred to the excellent review by Keys (1980) for a more in-depth discussion of glycine–serine interconversion.

Serine production

Since serine hydroxymethyltransferase (SHMT) is present in most plant tissues, it is possible that some serine could be made from glycine in non-green tissues in the same way as occurs in leaves during photorespiration. However, except under some special circumstances (see below), it is unlikely that glycine is available in these tissues except as a product of SHMT acting in the opposite direction, producing glycine from serine.

There are two other routes for serine synthesis (Fig. 1), both starting with 3'-phosphoglycerate which, depending on the tissue, may be derived from glycolysis or perhaps from the oxidative or reductive pentose phosphate pathways. The first of these two routes is commonly referred to as the phosphoserine pathway, and starts with the action of 3'-phosphoglycerate dehydrogenase which reduces the 3'-phosphoglycerate to O-phosphohydroxypyruvate, which is transaminated to O-phosphoserine, then dephosphorylated to serine. The second route starts with the dephosphorylation of 3'-phosphoglycerate to glycerate, by 3'-phosphoglycerate phosphatase, followed by reduction of the glycerate to hydroxypyruvate which is transaminated to serine (Keys, 1980).

Since there are two possible routes available to provide serine, the question arises as to which (if either) is used in different tissues. Walton and Woolhouse (1986) measured the levels of the enzymes of the serine synthesis pathways in leaves, germinating cotyledons, and root apices of pea plants. All the tissues examined contained high levels of phosphatases that could hydrolyse 3'-phosphoglycerate and O-phosphoserine, and the leaves contained all the other enzymes of the two pathways. The roots and germinating cotyledons, however, had extremely low levels of glycerate dehydrogenase. The authors suggested that the phosphoserine pathway was the principal route for serine synthesis in roots, while either pathway might play a role in serine synthesis in leaves during darkness. The germinating cotyledons had no detectable phosphohydroxypyruvate aminotransferase activity and hence no conclusions could be drawn regarding the route of serine synthesis in this tissue.

Glycine production

In germinating seeds that use fats or oils for storage, such as cotton or castor bean, acetyl CoA produced by the degradation of these

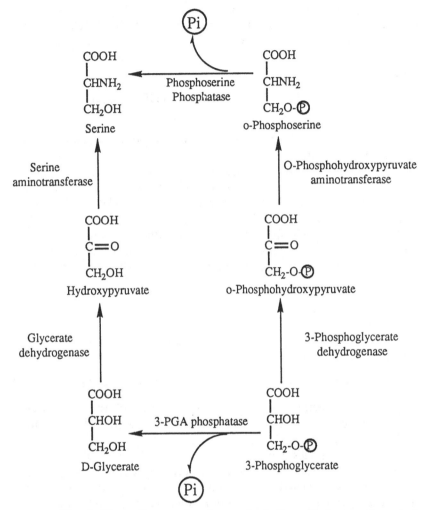

Fig. 1. The pathways of serine synthesis from 3'-phosphoglycerate in plant tissued. Adapted from Keys (1980).

storage reserves is fed into gluconeogenesis by the action of the glyoxylate cycle (Halpin & Lord, 1990). Acetyl CoA is converted to isocitrate (in the glyoxysome) which is then cleaved by isocitrate lyase to glyoxylate. So, for the few days that this pathway is operating in these tissues, there is a supply of glyoxylate that could serve as a substrate for the production of glycine by transamination. However, glycine synthesis in glyoxysomes has yet to be demonstrated.

Another possible route for glycine synthesis in plants is the reverse of the reaction catalysed by glycine decarboxylase. In vertebrate liver tissues (see Cossins, 1980) the enzyme, glycine synthase (E.C.2.1.2.10 – the same as glycine decarboxylase) assembles two one-carbon fragments into glycine:

$$CO_2 + NH_3 + NADH + H^+ + methyleneTHF \rightleftharpoons glycine + THF + NAD^+$$

This pathway has never been demonstrated to occur in plant tissues, but Sarojini and Oliver (1983) showed that glycine decarboxylase isolated from pea leaf mitochondria was reversible and could catalyse the synthesis of glycine from labelled bicarbonate. In roots, glycine decarboxylase levels are extremely low, only 2 to 5% of leaf levels (Walton and Woolhouse, 1986), and it is unlikely that this enzyme contributes in a significant way to glycine synthesis in this tissue.

In non-green tissues, most glycine is derived from serine by the action of SHMT, which cleaves serine to give glycine and methylene tetrahydrofolate.

Serine hydroxymethyltransferase

SHMT is a reversible enzyme with the equilibrium slightly in favour of glycine production.

$$serine + THF \rightleftharpoons glycine + N^5N^{10}\text{-methylene THF}$$

In leaves, the reaction is driven in the direction of serine synthesis by the high concentrations of glycine that are produced by photorespiration.

Root and leaf tissues contain similar levels of SHMT, but roots have very little glycine decarboxylase activity (Walton & Woolhouse, 1986; Rawsthorne, unpublished observations). This supports the notion that root SHMT plays a different role from its leaf counterpart, synthesizing glycine instead of consuming it. Since the SHMT involved in the photorespiratory pathway is closely linked with glycine decarboxylase, while that in roots seems to be free of this association, it would not be surprising to find the leaf and root enzymes exhibiting different properties.

One of the first studies on the kinetic properties of SHMT isolated from a non-photosynthetic tissue was on the enzyme from etiolated mung bean seedlings (Rao & Rao, 1980, 1982). These studies showed that SHMT activity varied during germination, with the total activity starting high, peaking at 36 hours and then decreasing. Specific activity began low, peaked at 72 hours, and then decreased. The enzyme was purified from 72-hour mung bean (*Vigna radiata* L.). seedlings and its native molecular weight determined to be 205 kD, consisting of four polypeptides, each of 50 kD. These values are similar to those obtained for SHMT from leaf tissues (e.g. Turner, Ireland & Rawsthorne 1992; Henricson & Ericson, 1988). The mung bean enzyme was found to have sigmoidal saturation kinetics for tetrahydrofolate, and this sigmoidicity was increased in the presence of NAD^+, and abolished in the presence of NADH. Glycine inhibited the enzyme when L-serine was the varied substrate. The authors concluded that SHMT is a regulatory enzyme in this tissue, controlling the interconversion of folate compounds and flux through subsequent one-carbon metabolism. Recently, Sukanya *et al.* (1991) have found that, unlike SHMT from other tissues, SHMT from germinating mung beans is not dependent on pyridoxal-5'-phosphate for activity. This requirement has been established for SHMT from animal (e.g. Fujioka, 1969) and some other plant (e.g. Prather & Sisler, 1966) sources, but further work on this is clearly needed.

Another non-photosynthetic tissue from which SHMT has been isolated is the soybean (*Glycine max* L.) nodule (Mitchell *et al.*, 1986). The nodule SHMT also had a molecular weight of 230 kD, and was again a tetramer consisting of four 55 kD peptides. In contrast to the findings for SHMT from mung bean seedlings described above, soybean nodule SHMT showed Michaelis–Menten kinetics with no evidence of cooperativity in substrate binding. In addition, NAD^+ and NADH had no effect on the kinetics of this enzyme, but glycine still acted as a competitive inhibitor. This suggests that the SHMT of soybean root nodules does not have the regulatory function suggested to exist in the etiolated mung bean seedling.

Several subcellular locations have been reported for SHMT. In green tissues most reports place it in the mitochondria (e.g. Turner *et al.*, 1992; Woo, 1979; Gardestrom *et al.*, 1985) but there are also reports of it being found in chloroplasts (Shah & Cossins, 1970; Shingles, Woodrow & Grodzinski, 1984). In non-green tissues it has been found in the cytosol (Rao & Rao, 1982) and in the plastid (Walton & Woolhouse, 1986).

Photosynthetic and non-photosynthetic forms of SHMT have similar pH optima, around pH 8.5 (e.g. Woo, 1979; Henricson & Ericson, 1988; Rao & Rao, 1982), but, as Walton and Woolhouse (1986) predicted, we now know there are also considerable differences between the enzymes isolated from these different tissues. A recent study by Turner *et al.* (1992) used an antibody raised against pea leaf mitochondrial SHMT to immunoprecipitate the SHMT activity in extracts of leaves and roots. The antibody was able to completely remove SHMT activity from mitochondrial extracts, but when used against whole leaf extracts, complete removal of activity could not be achieved – there was always 20–30% of the activity that could not be removed, regardless of how much antibody was added, suggesting the presence of another non-mitochondrial form of the enzyme that was not recognized by the antibody. When used against root extracts, the antibody was barely able to remove any activity, indicating the presence of an immunologically distinct form of the enzyme in roots.

Turner et al. (1992) found that mitochondrial extracts from leaf tissue yielded two distinct peaks of SHMT activity when fractionated using ion exchange (FPLC) chromatography. These two enzymes were immunologically indistinguishable from one another and as yet it is unclear what roles they may play in the pathways of the tissue. When whole leaf extracts were fractionated in the same way, a third peak of activity was found. The levels of the three forms varied during the development of the leaf (Ireland, unpublished observations). The presence of these multiple forms suggests that the different isoforms of SHMT may be responsible for the different functions of the enzyme in leaf tissue. Perhaps one form is present to handle the enormous flux through the photorespiratory pathway while another is responsible for glycine synthesis via the reverse pathway in the dark.

In contrast, when root extracts were fractionated in this way, only a single peak of SHMT activity was found. This is consistent with a need for only one form of the enzyme in roots since only the glycine synthesis pathway appears to be active.

We have determined the molecular weight of the root form of SHMT to be 115 000, which is approximately half that of the native enzyme isolated from leaf or nodule tissue. The molecular weight of root SHMT is closer to the 95 000 reported as the molecular weight of SHMT isolated from E.coli (Schirch et al., 1985): this may be related to the plastid location of root SHMT reported by Walton and Woolhouse (1986).

Conclusion

Considerable progress has been made in recent years on the biochemistry and molecular biology of the enzymes of glycine–serine interconversion in leaves. However, we know far less about their non-photorespiratory metabolism in other tissues, or even in leaf tissue in the dark. The presence of multiple and immunologically distinct forms of SHMT in different tissues and subcellular locations is consistent with it playing several roles, in accordance with the requirements of different tissues, which need to be more clearly elucidated.

References

Cossins, E.(1980). One carbon metabolism. In *The Biochemistry of Plants* Vol 2, ed. D.D. Davies, pp. 365–418. New York, Academic Press.

Fujioka, M.(1969). Purification and properties of SHMT from soluble and mitochondrial fractions of rabbit liver. *Biochimica et Biophysica Acta*, **185**, 338–49.

Gardestrom, P., Edwards, G.E., Henricson, D. & Ericson, I. (1985). The location of serine hydroxymethyltransferase in leaves of C3 and C4 species. *Physiologia Plantarum*, **64**, 29–33.

Halpin, C. & Lord, M.(1990). The structure and formation of micro-bodies. In *Plant Physiology, Biochemistry and Molecular Biology*, ed. D.T.Dennis & D.H. Turpin, pp. 329–38. Longman.

Henricson, D. & Ericson, I.(1988). Serine hydroxymethyltransferase from spinach leaf mitochondria. Purification and characterization. *Physiologia Plantarum*, **74**, 602–6.

Keys, A.J.(1980). Synthesis and interconversion of glycine and serine. In *The Biochemistry of Plants* Vol 5, ed.P.K. Stumpf & E.E. Conn, pp. 359–74. New York, Academic Press.

Mitchell, M.K., Reynolds, P.H.S. & Blevins, D.G.(1986). Serine hydroxymethyltransferase from soybean root nodules. *Plant Physiology*, **81**, 553–7.

Prather, C.W. & Sisler, E.C.(1966). Purification and properties of serine hydroxymethyltransferase from *Nicotiana rustica* L. *Plant and Cell Physiology*, **7**, 457–64.

Rao, D.N. & Rao, N.A.(1980). Allosteric regulation of serine hydroxymethyltransferase from mung bean (*Phaseolus aureus*). *Biochemical and Biophysical Research Communications*, **92**, 1166–71.

Rao, D.N. & Rao, N.A.(1982). Purification and regulatory properties of mung bean (*Vigna radiata* L.) serine hydroxymethyltransferase. *Plant Physiology*, **69**, 11–18.

Rhodes, D. & Hanson, A.D.(1993). Quaternary ammonium and tertiary sulfonium compounds in higher plants. *Annual Reviews of Plant Physiology*, **44**, 357–84

Sarojini, G. & Oliver, D.J.(1983). Extraction and partial characterization of the glycine decarboxylase multienzyme complex from pea leaf mitochondria. *Plant Physiology*, **72**, 194–9.

Schirch, V., Hopkins, S., Villar, E. & Angelaccio, S. (1985). Serine hydroxymethyltransferase from *Escherichia coli*: purification and properties. *Journal of Bacteriology*, **163**, 1–7.

Shah, S.P.J. & Cossins, E.A. (1970). The biosynthesis of glycine and serine by isolated chloroplasts. *Phytochemistry*, **9**, 1545–51.

Shingles, R., Woodrow, L., & Grodzinski, B. (1984). Effects of glycolate pathway intermediates on glycine decarboxylation and serine synthesis in pea (*Pisum sativum* L.). *Plant Physiology*, **74**, 705–10.

Sukanya, N., Vijaya, M., Savithri, H.S., Radhakrishnan, A.N. & Rao, N.A. (1991). Serine hydroxymethyltransferase from mung bean (*Vigna radiata*) is not a pyridoxal-5'-phosphate – dependent enzyme. *Plant Physiology*, **95**, 351–7.

Turner, S.R., Ireland, R.J. & Rawsthorne, S. (1992). Identification and localisation of multiple forms of serine hydroxymethyltransferase in pea (*Pisum sativum*) and characterisation of a cDNA encoding a mitochondrial isoform. *Journal of Biological Chemistry*, **267**, 13528–34.

Walton, N.J. & Woolhouse, H.W. (1986). Enzymes of serine and glycine metabolism in leaves and non-photosynthetic tissues of *Pisum sativum* L. *Planta*, **167**, 119–28.

Woo, K.C. (1979). Properties and intramitochondrial localisation of serine hydroxymethyltransferase in leaves of higher plants. *Plant Physiology*, **63**, 783–7.

E.G. BROWN

Biogenesis of N-heterocyclic amino acids by plants: mechanisms of biological significance

Introduction

The N-heterocyclic amino acids of plants mostly comprise a group of β-substituted alanines of which histidine (I) and tryptophan (II) are commonly occurring examples. Both of these compounds are found in a free state and as constituents of proteins. Biosynthetically, histidine derives from ATP and 5-phosphoribosyl pyrophosphate (Fig. 1) in a series of reactions in which the pyrimidine ring of the purine moiety of ATP is opened to release 5-amino-4-imidazole carboxamide ribotide,

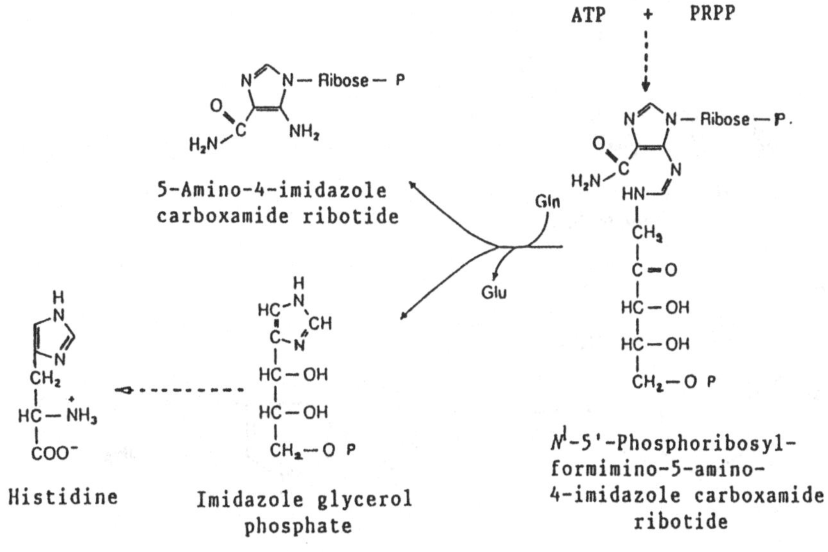

Fig. 1. Biosynthesis of histidine from ATP and 5-phosphoribosyl-1-pyrophosphate (PRPP). The byproduct 5-amino-4-imidazole carboxamide ribotide is an intermediate in purine biosynthesis.

an intermediate in the purine biosynthetic sequence. The residual N-atom and C-atom from the pyrimidine ring, together with the ribose moiety of the nucleotide, and an additional N-atom donated by gluta-mine, form a new imidazole ring with a 3C side chain, eventually yielding histidine. The pathway involved is well documented and the enzymes have been subjected to close scrutiny. Tryptophan has its origins, via the shikimate pathway, in indole 3-glycerol phosphate. The enzyme tryptophan synthase converts this to tryptophan using serine as the donor of an intact alanine side-chain (Fig. 2a). As with histidine, the biosynthetic route to tryptophan is well established and docu-mented. However, the tryptophan synthase reaction is also of interest

(a)

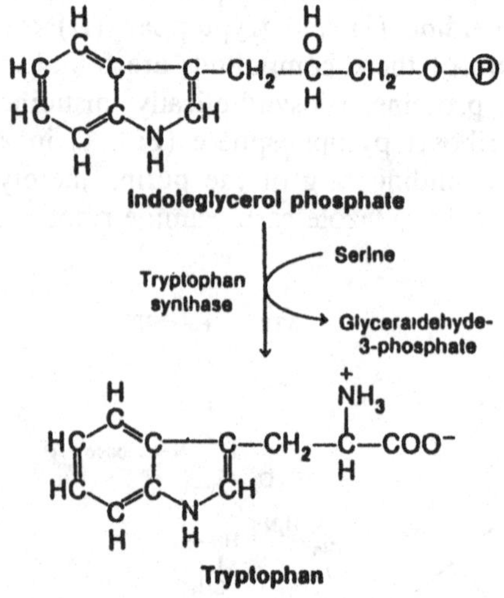

Indoleglycerol phosphate

Tryptophan
synthase

Serine

Glyceraldehyde-
3-phosphate

Tryptophan

(b)

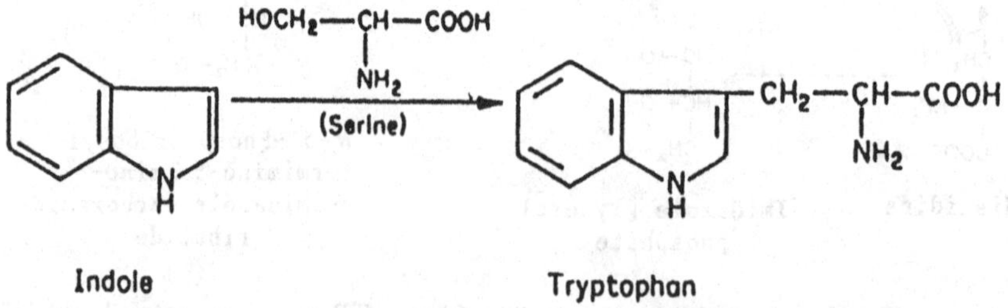

Indole

Tryptophan

Fig. 2. Reaction catalysed by **a** tryptophan synthase (holoenzyme), and by **b** the isolated β-subunits of tryptophan synthase.

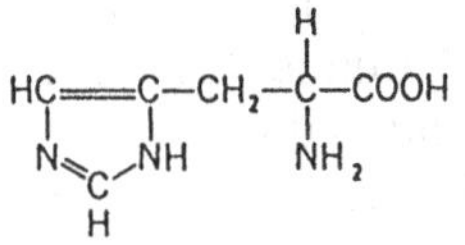

L-Histidine

I

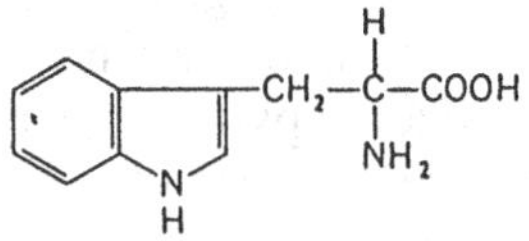

L-Tryptophan

II

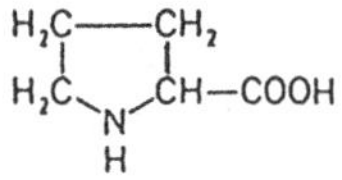

L-Proline

III

in relation to the formation of some of the less common, so-called 'non-protein' amino acids of plants with which this account is primarily concerned. The enzyme has been isolated from various plant, animal and microbial sources, and has been the subject of extensive research. It is dissociable into two α-subunits and two β-subunits and whereas the holoenzyme catalyses the reaction described in Fig. 2a, the isolated β-subunits catalyse a simpler reaction in which indole and serine are condensed to form tryptophan (Fig. 2b). The fascinating aspect of the mechanism of tryptophan synthase is that it proceeds in two steps, the first of which is catalysed by the α-subunits, releasing indole from indol-3-ylglycerol phosphate and channelling it through an interior tunnel within the protein structure to the β-subunits (Hyde *et al.*, 1988). The indole itself never leaves the enzyme in a free state.

Proline (III), although an imino acid, is a component of the primary structure of proteins, and is therefore conventionally included in discussion of N-heterocyclic amino acid biochemistry. In plants, it is formed by the spontaneous cyclization of glutamic 5-semialdehyde, followed by enzymic reduction of the resulting pyrroline ring (Noguchi, Koiwai & Tamaki, 1966; Fig. 3).

In addition to the common amino acid constituents of proteins, plants produce a number of N-heterocyclic 'non-protein' amino acids. Most

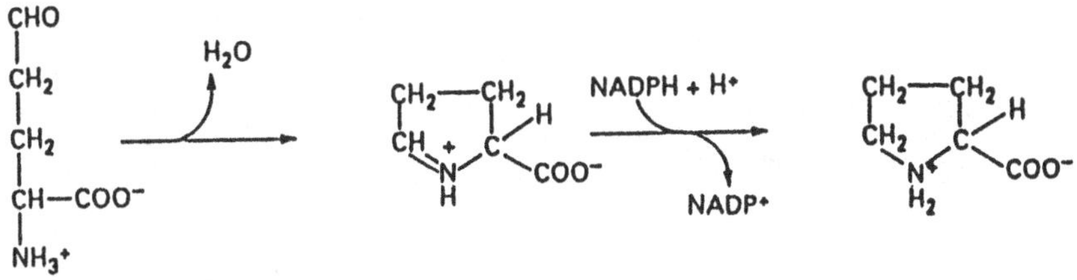

glutamic 5-semialdehyde

Δ¹-pyrroline 5-carboxylic acid

proline

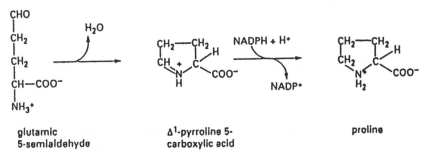

Fig. 3. Biosynthesis of proline in higher plants.

of the recent studies on the biogenesis of these compounds have concerned the isomeric pyrimidine amino acids willardiine (IV) and isowillardiine (V), the related pyrimidine amino acid lathyrine (VI) and the pyrazole derivative β-pyrazol-1-ylalanine (VII). It is the biochemistry of these N-heterocyclic β-substituted alanines with which the remainder of this account deals.

General mechanisms of formation of N-heterocyclic β-substituted alanines

In each case, the biosynthesis of the pyrimidine amino acids and of β-pyrazol-1-ylalanine results in the separate formation of the N-heterocyclic ring and its subsequent reaction with serine or a serine derivative to form the alanine side chain. In this respect, their origin has parallels with the formation of the amino acid tryptophan (Fig. 2a, b) and especially the reaction catalysed by the isolated β-subunits of tryptophan synthase (Fig. 2b). For a time, it was therefore thought likely that a plant tryptophan synthase of low specificity was responsible for the formation of these analogues (Dunnill & Fowden, 1963; Hadwiger et al., 1965; Tivari, Penrose & Spenser, 1967). Although, both on theoretical grounds and on an experimental basis, it was subsequently considered

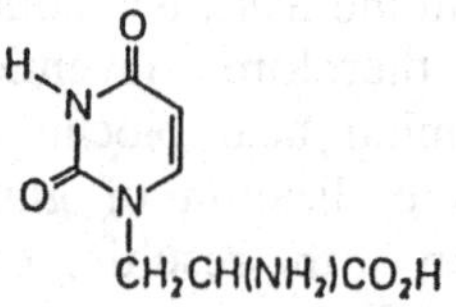

Willardiine
β-(2,4-dihydroxypyrimidin-1-yl)alanine

IV

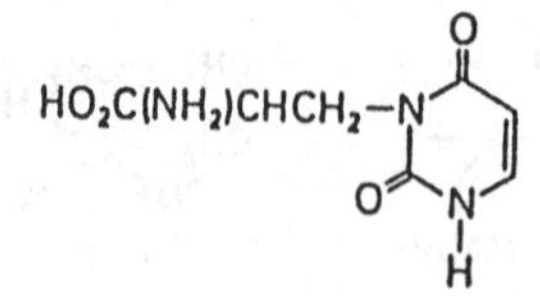

Isowillardiine
β-(2,4-dihydroxypyrimidin-3-yl)alanine

V

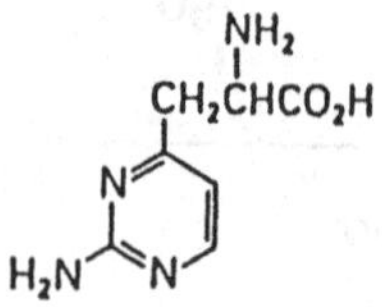

Lathyrine
β-(2-aminopyrimidin-4-yl)alanine

VI

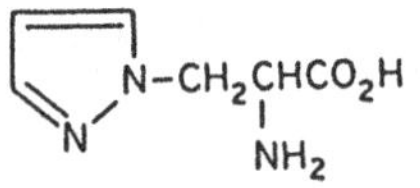

β-pyrazol-1-ylalanine

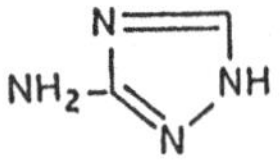

Amitrol (3-amino-1,2,4-triazole)

VII

VIII

unlikely that willardiine and isowillardiine arise from the activity of tryptophan synthase (Brown, 1970), it was nevertheless shown by isotopic tracer studies that these two amino acids are synthesized from a preformed N-heterocyclic base, namely uracil, and an alanine donor (Ashworth, Brown & Roberts, 1972). Following this report, attention then began to focus on another enzymic possibility for coupling the base and the alanine donor, this centred on the activity of the enzyme reported by Dunnill and Fowden (1963) to catalyse formation of β-pyrazol-lylalanine from the N-heterocycle pyrazole and serine (Fig. 4). This enzyme, β-pyrazol-l-ylalanine synthase, had been shown by Fowden and his coworkers to have a relatively wide specificity and to catalyse formation of a group of N-heterocyclic β-substituted alanines from a number of naturally occurring and synthetic N-heterocycles (Frisch et al., 1967). Compounds such as triazoles, including the herbicide Amitrol (3-amino-1,2,4-triazole) (VIII) and tetrazoles, all served as substrates. Whereas the relatively wide specificity of β-pyrazol-l-ylalanine synthase was amply confirmed by subsequent work from Fowden's group and by others, the enzyme is nevertheless incapable of catalysing formation of either willardiine or isowillardiine from uracil (Ahmmad, Maskall & Brown, 1984).

Later studies with extracts of watermelon seedlings containing β-pyrazolylalanine synthase activity (Murakoshi, Kuramoto & Haginawa, 1972) showed that the real alanine side chain donor is not in fact serine but O-acetylserine, a compound which Giovanelli and Mudd (1967) had shown to be the precursor of cysteine (Fig. 5). This was

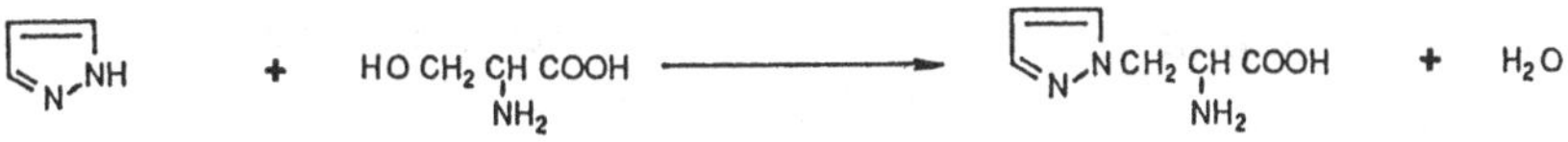

Fig. 4. Condensation of pyrazole and serine to yield β-pyrazol-l-ylalanine.

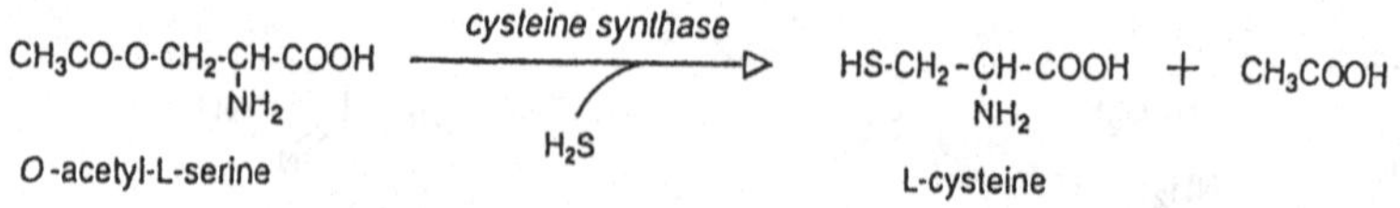
cysteine synthase
CH3CO-O-CH2-CH-COOH
NH2
O-acetyl-L-serine
H2S
HS-CH2-CH-COOH + CH3COOH
NH2
L-cysteine

Fig. 5. Enzymic formation of cysteine from O-acetylserine and H_2S.

then demonstrated to be also the case in willardiine and isowillardiine formation (Ashworth *et al.*, 1972). The reason why serine had appeared to be the alanine donor originally was because the crude enzyme preparations being used at that time were capable of acetylating it, through the agency of acetyl-CoA and the enzyme L-serine acetyltransferase (Ngo & Shargool, 1974).

Recently the concept of an enzymic mechanism of low specificity being involved in the formation of N-heterocyclic β-substituted alanines has reappeared and now implicates cysteine synthase. Ikegami *et al.* (1987) presented evidence that some cysteine synthases from higher plants can also catalyse the formation of β-substituted alanines such as mimosine (IX), quisqualic acid (X), and β-(isoxazolin-5-on-2-yl)alanine (XI) but once again willardiine and isowillardiine are not produced by these low specificity enzymic systems (Ikegami *et al.*, 1987).

Special mechanisms of formation

The formation of β-pyrazol-l-ylalanine, mimosine, willardiine and isowillardiine, and other N-heterocyclic β-substituted alanines, by transfer of an alanine residue from O-acetylserine to the appropriate hetero-

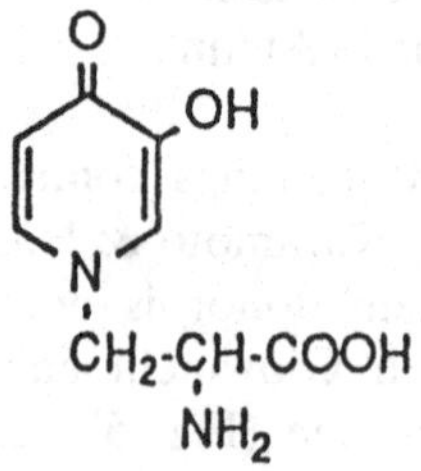
O
OH
N
CH2-CH-COOH
NH2

L-mimosine

IX

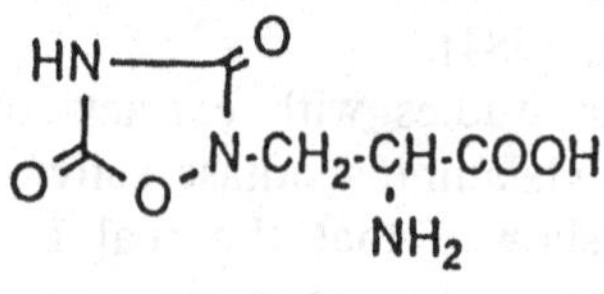
HN
O
N-CH2-CH-COOH
NH2

L-quisqualic acid

X

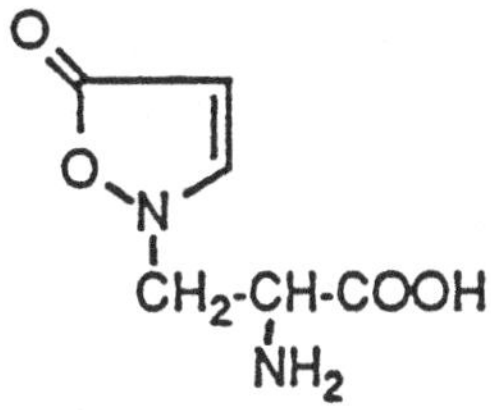

XI

cyclic base is now well established. In some cases the general enzymic mechanism with cysteine synthase appears to operate but in others, especially those involving the pyrimidine amino acids, the enzymes are specific. Interestingly, the formation of 3-(3-amino-3-carboxy-propyl)uridine (XII) which is the riboside of the next higher homologue of isowillardiine (Ohashi *et al.*, 1974) does not use O-acetylhomoserine as the side chain donor as might have been expected but involves the transfer to uridine of a 3-amino-3-carboxypropyl group from S-adenosyl-methionine (XIII) (Nishimura *et al.*, 1974).

Biosynthesis of the pyrimidine amino acid lathyrine also presents an example of a special mechanism for the biosynthesis of a N-heterocyclic β-substituted alanine. Lathyrine (VI) differs from the other two pyr-

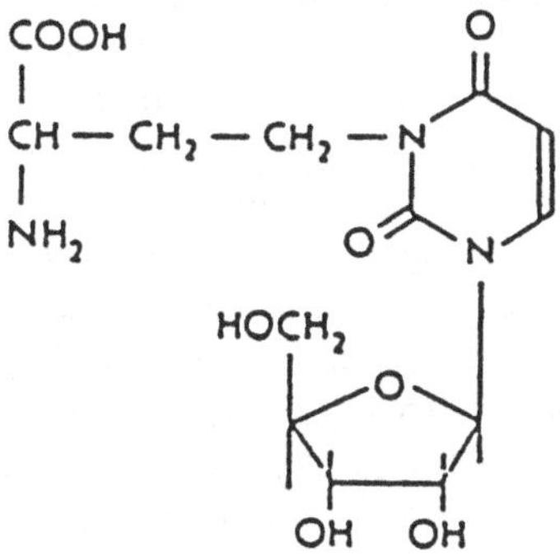

XII

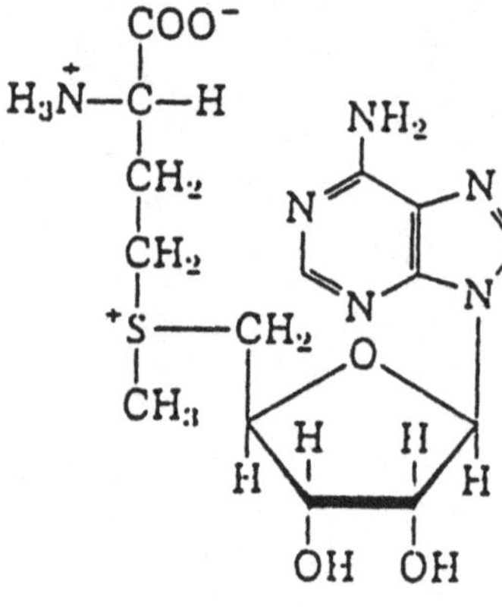

S-Adenosylmethionine

XIII

imidine amino acids willardiine (IV) and isowillardiine (V) in that the alanine side chain is attached to a ring C-atom and not to an N-atom. Bell (1963), Nowacki and Nowacka (1963) and Bell and Przybylska (1965) had postulated that lathyrine originates from homoarginine through the cyclization of 4-hydroxyhomoarginine (Fig. 6). The evidence for this was that radioactivity from ^{14}C-homoarginine is incorporated into lathyrine by seedlings of *Lathyrus tingitanus* and that 4-hydroxyhomoarginine and homoarginine can both be detected as natural constituents of these seedlings. However, during a comparative study in this laboratory of pyrimidine metabolism by various legume seedlings, it was observed that there was a substantial incorporation of radioactivity into lathyrine from ^{14}C-orotate when the latter was supplied to seedlings of *L. tingitanus* (Brown, 1975). As orotate is a key intermediate in the assembly of most naturally occurring pyrimidine molecules, it was suspected that despite the homoarginine findings, the biosynthesis of lathyrine was no exception and that its pyrimidine moiety was synthesized via the orotate pathway of pyrimidine biosynthesis. Subsequent studies (Brown & Al-Baldawi, 1977; Al-Baldawi & Brown, 1983) confirmed the substantial incorporation of radioactivity from [6-^{14}C]orotate into lathyrine and further showed incorporation

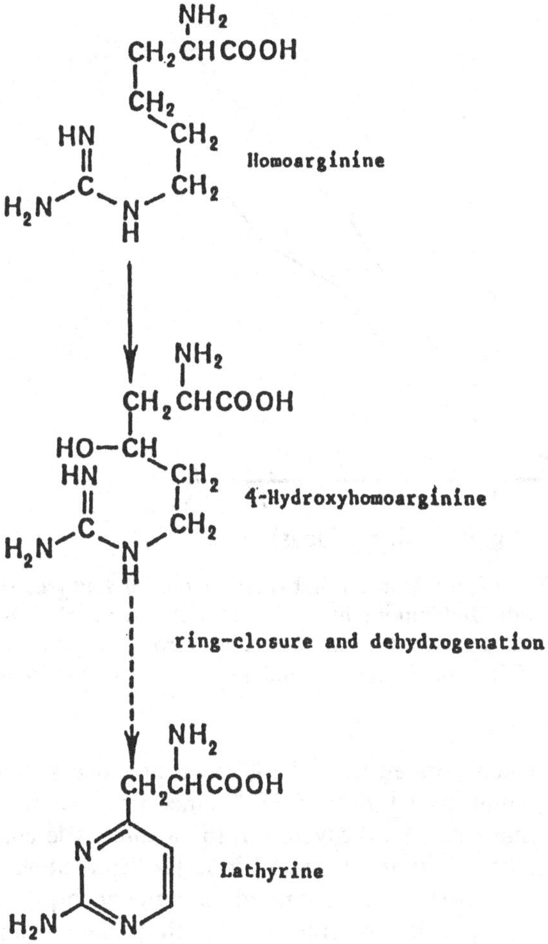

Fig. 6. Mechanism originally proposed for the biosynthesis of lathyrine from 4-hydroxyhomoarginine.

of [2-^{14}C]uracil and [3-^{14}C]serine into this N-heterocyclic amino acid. Simultaneous presence of a pool of non-radioactive uracil, supplied exogenously, was observed to decrease substantially the incorporation of radioactivity from [6-^{14}C]orotate into lathyrine by *L.tingitanus*. As it could be demonstrated that uracil does not inhibit lathyrine biosynthesis, but to the contrary stimulates it (Fig. 7), the observation implies that in the biosynthetic sequence uracil is closer to lathyrine than is orotate. Systematic chemical degradation of lathyrine to locate the

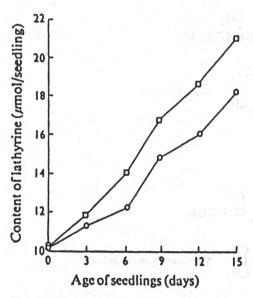

Fig. 7. Effect of uracil on the lathyrine content of *Lathyrus* seedlings during growth. Imbibition, by seeds, of water (controls; O—O) or 50 mM uracil solution (□—□) began at zero time (Brown & Al-Baldawi, 1977; reproduced by kind permission of the *Biochemical Journal*).

position of the ^{14}C incorporated from the labelled precursors (Brown & Al-Baldawi, 1977) showed that 90% of the ^{14}C incorporated into lathyrine from [3-^{14}C]serine could be recovered in the alanine side chain and over 80% of the radioactivity incorporated from [2-^{14}C]uracil was found in C-2 of lathyrine. These data confirmed that the incorporation of uracil is direct and specific, in other words that uracil supplies a preformed pyrimidine ring for lathyrine biosynthesis, and that serine donates an intact 3-C unit to form the alanine side chain. Paradoxically, however, it was also possible to confirm Bell's finding that some radioactivity from [guanidino^{14}C]homoarginine is incorporated into lathyrine but the extent of this was found to be relatively insignificant, less than 4% of that incorporated from a similar amount of [2-^{14}C]-uracil of comparable specific radioactivity (Brown & Al-Baldawi, 1977). Consideration of the pathway of pyrimidine catabolism (Fig. 8) led Brown and Al-Baldawi (1977) to the view that if lathyrine passed along a similar degradative route to that followed by uracil and 5-methyluracil (thymine) the first two products would be 5,6-dihydrolathyrine and 4-hydroxyhomoarginine. The small amount of radioactivity apparently

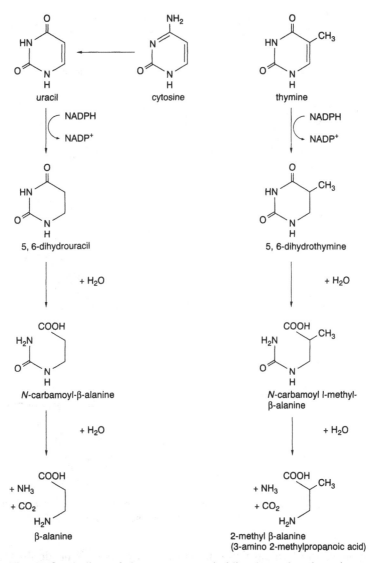

Fig. 8. Catabolism of the common pyrimidine bases by plant tissues.

incorporated into lathyrine from [guanidino-^{14}C]homoarginine could therefore result from a minor reversal of the first two steps of the catabolic sequence induced by the mass action effect of exogenously supplied homoarginine. This would only involve a biochemically facile dehydrogenation and condensation (Fig. 9). Whether this is the expla-

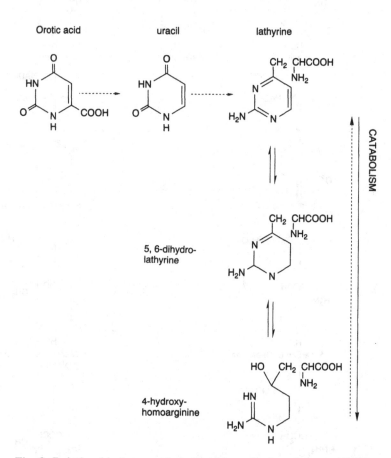

Fig. 9. Relationship between lathyrine biosynthesis and the catabolism of lathyrine to 4-hydroxyhomoarginine (Brown & Al-Baldawi, 1977).

nation of the homoarginine findings or not, there is no doubt that the primary route to lathyrine production is the orotate pathway of pyrimidine biosynthesis and not cyclization of 4-hydroxyhomoarginine.

Despite the foregoing experimental clarification of lathyrine biosynthesis, demonstrating that this compound originates from a preformed pyrimidine ring coming from the orotate pathway, a major problem still remained. Elucidation of the general biochemical mechanism for formation of N-heterocyclic β-substituted alanines from an appropriate

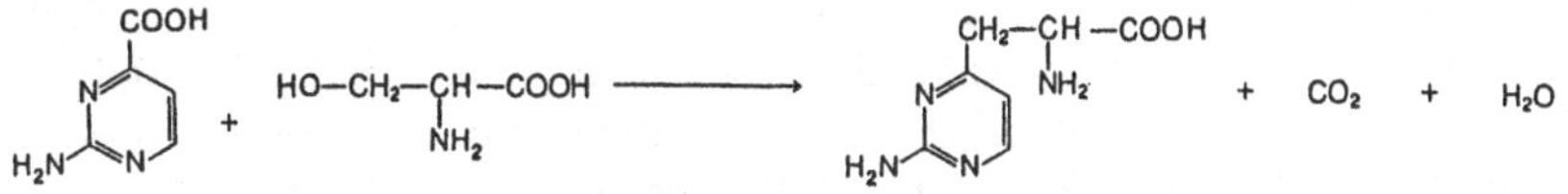

Fig. 10. Reaction catalysed by lathyrine synthase during lathyrine biosynthesis by *Lathyrus tingitanus* (Brown & Mohamad, 1990).

heterocyclic base and O-acetylserine, discussed above, suggested that formation of lathyrine involved a similar process. However, using both *in vivo* and *in vitro* methods, exhaustive testing of possible pyrimidine precursors was unsuccessful; compounds such as 2-aminopyrimidine and 2-amino-4-hydroxypyrimidine (isocytosine) would not act as lathyrine precursors in seedlings, nor as alanine acceptors for the formation of lathyrine by enzymic extracts of *Lathyrus* and other plant species, neither did they dilute the specific radioactivity of lathyrine synthesised from ^{14}C-uracil in *Lathyrus* seedlings in the presence of large pools of these compounds (Brown & Al-Baldawi, 1977). Later work in this laboratory (Brown & Mohamad, 1990) eventually identified the elusive precursor as 2-amino-4-carboxypyrimidine and showed that it serves as an acceptor of an alanine residue from serine in an enzyme-catalysed reaction in which the 4-carboxy group is eliminated as CO_2 (Fig. 10). The enzyme, from *Lathyrus* seedlings, was partially purified and shown to be pyridoxal 5-phosphate dependent (Brown & Mohamad, 1990). Curiously, and unlike the other enzymes catalysing formation of N-heterocyclic β-substituted alanines, it would only use serine and not O-acetylserine as the alanine donor. The reaction parallels one described earlier by Hadwiger *et al.* (1965) in which orcylalanine is formed from orsellinic acid and serine with concomitant elimination of the ring carboxyl group as CO_2 (Fig. 11). The lathyrine-forming enzyme has a requirement for biotin, an unusual requirement for a biological decarboxylation enzyme as distinct from one involved in carboxylation. This requirement was confirmed not only by the observed seven-fold stimulation of the dialysed enzymic activity by biotin (Brown & Mohamad, 1990) but by the inhibition of activity seen in the presence of avidin.

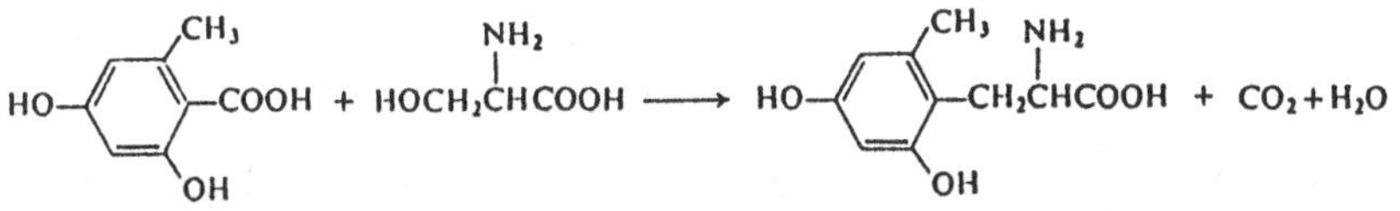

Fig. 11. Enzymic formation of orcylalanine from orsellinic acid and serine (Hadwiger *et al.*, 1965).

The biogenesis of tryptophan, mimosine, β-pyrazol-l-ylalanine and other N-heterocyclic β-substituted alanines can be successfully modelled in non-enzymic reactions involving pyridoxal and appropriate di- or trivalent metal ions as catalysts; Al^{3+} ions have been commonly used in such studies. These pyridoxal model systems stem from the observations of Metzler, Ikawa and Snell (1953) on the formation of tryptophan from serine and indole in heated, nonenzymic, buffered solutions also containing pyridoxal and $Al_2(SO_4)_3$, and their demonstration that this approach could be used to model a number of pyridoxal phosphate-dependent enzymic reactions. For those working in the field of heterocyclic β-substituted alanine biochemistry, it has almost become *de rigueur* to demonstrate not only that the compound being studied can be formed biologically but that it can be produced synthetically in a pyridoxal model system. In practice, this affords a useful and simple feasibility test for hypotheses concerning the origin of novel heterocyclic β-substituted alanines. It has also proved to be invaluable in the laboratory for the small-scale synthesis of commercially unavailable amino acids of the type being considered.

Whereas the pyridoxal model system has worked well with many of the N-heterocyclic β-substituted alanines that occur in nature, it proved difficult to apply to others. For example, the pyrimidine amino acids willardiine and isowillardiine proved to be difficult to produce in this way. The key to eventual success with willardiine was the finding (Murakoshi *et al.*, 1978) that gallium ions (Ga^{3+}) were considerably more effective than Al^{3+} ions but it was reported that the formation of isowillardiine at 35° for 2–3 h was negligible. Ahmmad, Maskell and Brown (1984) subsequently found that by raising the temperature to 100°C, for 30 min, a significant amount of isowillardiine was produced. Examination of other group IIIA elements by Ahmmad *et al.* (1984) showed that, at 100°C, gallium ions were the most effective and that indium ions were virtually devoid of activity; borate and aluminium ions were of intermediate efficiency. Once lathyrine had been shown to be formed biologically from 2-amino-4-carboxypyrimidine and serine (Brown & Mohamad, 1990) this reaction was also tested in the pyridoxal model system. As with willardiine and isowillardiine formation, it was found that Ga^{3+} ions were effective whereas Al^{3+} ions were not.

Biosynthesis of the heterocyclic moiety

So far, this account of the biogenesis of the N-heterocyclic β-substituted alanines has concentrated on their formation in pyridoxal phosphate dependent enzymic reactions from appropriate N-heterocyclic bases and either serine or O-acetylserine as donors of the alanine side chain. This is, however, only half of the story and no account of their biosynthesis would be complete without consideration of the origin of the individual, component heterocyclic bases. Indeed, 25 years ago Fowden (1967) had written that the production of β-pyrazol-l-ylalanine by certain cucurbits rests upon their unique ability to elaborate the pyrazole ring and not their possession of a specific condensing enzyme.

With the pyrimidine amino acids willardiine and isowillardiine, the question is, then, what is the source of the uracil from which they are formed? As would be expected from what is known of pyrimidine metabolism, the uracil originates from the orotate pathway. This was demonstrated by the substantial incorporation of radioactivity from [6-¹⁴C]-orotate into willardiine and isowillardiine by pea seedlings (Ashworth, Brown & Roberts, 1972), and by the observation in the same study that the labelling from [6-¹⁴C]-orotate was substantially diluted out in the presence of an exogenously supplied pool of unlabelled uracil and to a lesser extent by an exogenously supplied pool of either uridine or UMP. As uracil, supplied in this way, stimulates willardiine and isowillardiine synthesis (Fig. 12) and is certainly not an inhibitor, the data indicate that the free uracil used for willardiine and isowillardiine synthesis arises at a later stage in the biosynthetic sequence than these other uracil derivatives. Again, this is what would

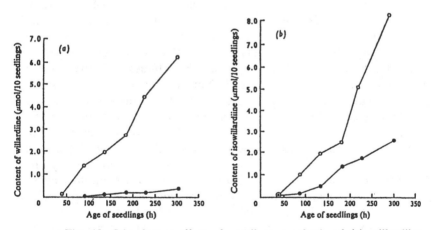

Fig. 12. Stimulatory effect of uracil on synthesis of (*a*) willardiine, and (*b*) isowillardiine by pea seedlings. Seeds were imbibed at zero time in either water (controls; ●—●). or 30 mM uracil solution (O—O) (Ashworth, Brown & Roberts, 1972; reproduced by kind permission of the *Biochemical Journal*).

be expected from the known pathway of pyrimidine biosynthesis which culminates in the formation of UMP, the key compound from which all the other commonly occurring pyrimidines are produced.

In studies of lathyrine biosynthesis using seedlings of Lathyrus tingitanus (Brown & Al-Baldawi, 1977), uracil was shown to stimulate production of this pyrimidine amino acid (Fig. 7) and ^{14}C-uracil was more efficiently incorporated into it than was ^{14}C-orotate. Further, the additional presence of an exogenously introduced pool of uracil substantially diluted the specific radioactivity of the lathyrine produced from ^{14}C-orotate although, clearly from Fig. 7, uracil stimulates and does not inhibit synthesis of the pyrimidine amino acid. In the converse experiment, exogenous provision of a pool of unlabelled orotate did not affect the specific radioactivity of the lathyrine synthesized from ^{14}C-uracil. It was concluded therefore that with lathyrine too, uracil is the more immediate precursor of the pyrimidine ring than is orotate (Brown & Al-Baldawi, 1977). The origin of the pyrimidine ring of willardiine, isowillardiine and lathyrine is summarized in Fig. 13, from which it can be seen that the major remaining problem is the mechanism of conversion of uracil to 2-amino-4-carboxypyrimidine. At least two enzymic steps must be involved in aminating and carboxylating uracil but at present nothing is known of these processes or of the enzymes that are undoubtedly involved.

Another of the N-heterocyclic β-substituted alanines that has been examined in the wider biosynthetic context of the origin of the heterocyclic base, is β-pyrazol-l-ylalanine. As was discussed above, this amino acid is formed in various cucurbits from pyrazole and O-acetylserine by a pyridoxal 5-phosphate dependent synthase (Dunnill & Fowden, 1963; Murakoshi et al., 1972) but prior to the studies of Brown, Flayeh and Gallon (1982) and Brown and Diffin (1990) nothing was known of the origin of the pyrazole needed for the process. In fact, the biosynthesis of pyrazole poses a fundamentally important biochemical question, namely, how do two N-atoms become covalently bound in nature? Experimental examination of possible precursors implicated 1,3-diaminopropane in the process and this was supported by the demonstration that [3,4-^{14}C]1,3-diaminopropane was significantly incorporated by cucumber seedlings into the pyrazole moiety of β-pyrazol-l-ylalanine (Brown, Flayeh & Gallon, 1982). Flayeh et al. (1984) subsequently showed that 1,3-diaminopropane is a natural constituent of seeds and seedlings of Cucumis sativus. More recent investigations (Brown & Diffin, 1990) have confirmed that 1,3-diaminopropane is the precursor of pyrazole and during this study, an enzymic extract was obtained from seedlings of C.sativus that catalysed both cyclization of

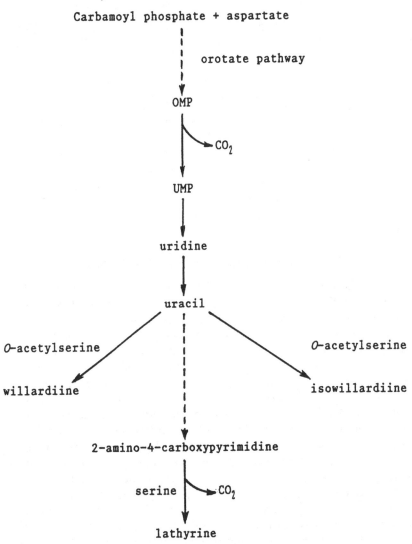

Fig. 13. Origin of the pyrimidine rings of the N-heterocyclic β-substituted alanines, willardiine, isowillardiine, and lathyrine.

the diamine and its subsequent dehydrogenation, according to the scheme shown in Fig. 14, to yield pyrazole. The enzymic activity was partially purified by fractional precipitation with ammonium sulphate and by gel-filtration chromatography on Sephadex G-100, and shown to be FAD-dependent. It was also confirmed during Brown and Diffin's

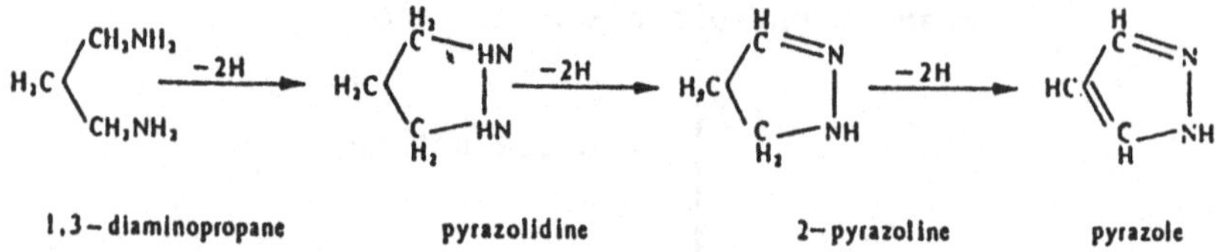

1,3–diaminopropane   pyrazolidine   2–pyrazoline   pyrazole

Fig. 14. Formation of pyrazole from 1,3-diaminopropane by cyclization and sequential dehydrogenation (Brown & Diffin, 1990; reproduced by kind permission of Pergamon Press).

study that 2-pyrazoline (Fig. 14) is an intermediate in the conversion of l,3-diaminopropane to pyrazole. The enzyme preparation is the first to be described, from any biological source, that catalyses the covalent linking of two organic N-atoms.

Possible functions and unanswered questions

Biosynthesis of the N-heterocyclic β-substituted alanines has proved to be a particularly interesting area of plant biochemistry and has resulted in the discovery of several new enzymic mechanisms. The presence of others is indicated but not yet investigated. Of particular importance for the future however is the question of the interaction of what would be currently perceived to be the pathways of primary metabolism with those of secondary metabolism. Contemplation of this question overlaps with another – does the production of N-heterocyclic β-substituted alanines, and indeed of other types of non-protein amino acid, serve any useful function for the producer plants? Over the years, there have been various speculations made about this. Some of the syntheses have been attributed to vestigial metabolic pathways, the original functions of which have been long-since extinct. Others have been implicated as defence mechanisms producing compounds that are toxic to, or deter attack by, predatory animals or invasive microorganisms. The suggestion that they arise as accidents of metabolism because of the low specificity of some plant enzymes is now less favoured since many of the enzymes studied in connection with the biosyntheses considered above, are highly specific. For example, although β-pyrazol-l-ylalanine synthase will form a number of unnatural analogues of β-pyrazol-l-ylalanine if supplied with appropriate heterocyclic bases (Frisch et al., 1967), it will not catalyse formation of the pyrimidine analogues willardiine and isowillardiine. Nitrogen storage is another suggested function of the N-heterocyclic amino acids that have been considered in this chapter. This is not an unattractive idea since relatively large amounts of some of these compounds can accumulate in plants. Lathyrine, for example, constitutes more than 2% of the dry weight of Lathyrus tingitanus seeds, and 20% of the total radioactivity incorporated into pyrimidines from [6-^{14}C]orotate by *Lathyrus* seedlings is recovered in lathyrine (Al-Baldawi & Brown, 1983). β-Pyrazol-l-ylalanine comprises about 0.1% of the total dry weight of watermelon seeds (Noe & Fowden, 1960). Further, in the case of β-pyrazol-l-ylalanine, it was shown (Brown, Flayeh & Gallon, 1982) that in germinating seeds it is released from the peptide γ-glutamyl-β-pyrazol-l-ylalanine during the first 10 days and further metabolized.

Another view of function, one which is particularly attractive, is that the N-heterocyclic β-substituted alanines are detoxification products. All living systems produce toxic N-containing wastes which they must void as expeditiously as possible. Most of these arise through protein and nucleic acid catabolism. In animals, compounds such as uric acid and urea are common excretory products. Plants, however, do not possess excretory systems as such, and in order to remove potentially hazardous compounds they are sequestered in vacuoles. The vacuoles, however, although of low general metabolic activity, are not totally inert and physiologically active compounds would need to be inactivated before being deposited in them. Furthermore, transporting potential metabolic poisons across a cell to a vacuole would be hazardous unless the compounds were first rendered harmless in some way. The production of β-pyrazol-l-ylalanine seems to offer a good case for plant detoxification. First of all, although there is little direct information available on toxic effects of pyrazole on plant tissues, it is undoubtedly toxic to various other biological systems. For example, it is a potent competitive inhibition of liver alcohol dehydrogenase (Theorell & Yonetani, 1963; Theorell, Yonetani & Sjoberg, 1969; Li & Theorell, 1969) and also inhibits the microsomal ethanol oxidizing system. Additionally, it inhibits catalase and the transport of reducing equivalents into the mitochondria (Cederbaum & Rubin, 1974). Chronic administration induces ultrastructural changes in rat liver (Lieber *et al.*, 1970) and affects brain noradrenaline concentrations (MacDonald, Marselas & Nousiainen, 1975). It affects enzymes, such as NADP-cytochrome c reductase, involved in hydroxylation processes (Marselos *et al.*, 1977 and also the enzymes of glucuronidation. UDP-glucose dehydrogenase, UDP-glucuronosyl transferase and L-gulonate dehydrogenase are markedly enhanced whereas the activities of UDP-glucuronic acid pyrophosphatase, β-glucoronidase, and D-glucuronolactone dehydrogenase are decreased (Marselos *et al.*, 1977).

Although Takeshita *et al.* (1963) reported the existence of an enzyme in *Pseudomonas* spp. that would hydrolyse β-pyrazol-l-ylalanine to pyrazole, pyruvic acid and ammonia (Fig. 15), no evidence for the

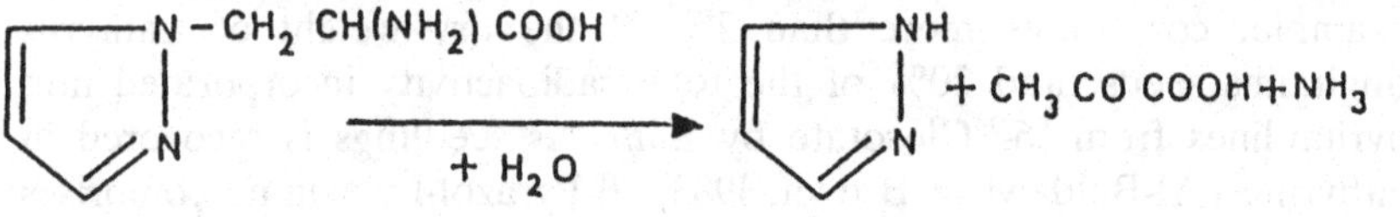

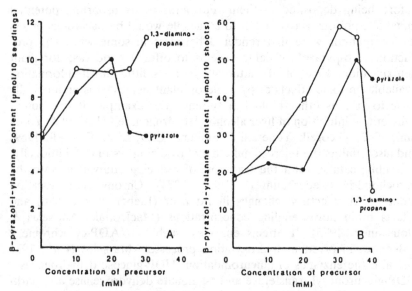

Fig. 15. Microbial hydrolysis of β-pyrazol-l-ylalanine (Takeshita, Nishi-zuka & Hayaishi, 1963).

existence of this enzyme in animal tissues was found by Al-Baldawi and Brown (1985) who also observed that 93% of the β-pyrazol-l-ylalanine fed to mice was recovered unchanged in the urine; no pyrazole could be detected. In summary then, pyrazole is toxic in animals but β-pyrazol-1-ylalanine is not. What about plants? Brown and Diffin (1990) showed that both pyrazole and its precursor 1,3-diaminopropane stimulated production of β-pyrazol-1-ylalanine by germinating seedlings and excised shoots of *Cucumis sativus* (Fig. 16a,b). Higher concentrations of either were, however, inhibitory, as would be expected if

Fig. 16. Effect of pyrazole (●—●) and of 1,3-diaminopropane (O—O) on the β-pyrazol-1-ylalanine content of (A) germinating seedlings of *Cucumis sativus*, and of (B) excised shoots of 23-day old seedlings (Brown & Diffin, 1990; reproduced by kind permission of Pergamon Press).

pyrazole is toxic to the plant and formation from it of β-pyrazol-l-ylalanine represents a detoxification mechanism. This is because the more pyrazole presented, the more β-pyrazol-l-ylalanine is made, until supply eventually exceeds the maximum rate at which the detoxification mechanism can operate. It is interesting that although 1,3-diamino-propane behaved similarly to pyrazole in the excised shoot experiment (Fig. 16b), higher concentrations of the diamine did not cause inhibition with germinating seeds (Fig. 16a). This suggests that 1,3-diamino-propane is not toxic *per se* but produces adverse effects if converted to pyrazole at a rate which causes pyrazole to accumulate and overload the detoxification mechanism. In the experiments shown in Fig. 16, the germinating seeds had at least 48 hr in which to metabolize the fixed amount with which they were presented during the imbibition phase but the excised shoots were taking up 1,3-diaminopropane con-tinuously throughout the 24 hr period before examination.

As described above, all three pyrimidine amino acids are formed from uracil. Willardiine and isowillardiine are formed directly, and lathyrine is formed from uracil via the intermediate synthesis of 2-amino-4-carboxypyrimidine. These processes may well represent uracil detoxification mechanisms. Certainly, in excised shoots of *Pisum sati-vum* and *Lathyrus tingitanus* all three mechanisms are able to cope with a constant exogenous supply of uracil (Figs. 7,12). Even concen-trations of uracil approaching its maximum solubility are metabolized without difficulty (Fig. 17). An interesting aspect of these uracil-metabolizing processes is that as the source of the uracil is the orotate pathway (Fig. 13) and because in plants, pyrimidine biosynthesis is regulated mainly through feedback inhibition by UMP on aspartate transcarbamoylase (Yon, 1971*a,b*), extensive conversion of UMP to uracil must simultaneously remove the feedback control. As all three pyrimidine amino acids are produced in quantity, this poses the question as to whether in pea seedlings there is overproduction of uracil stem-ming from excessive enzymic hydrolysis of UMP or whether pyrimidine biosynthesis is exaggerated by unidentified factor(s) and the ultimate product, uracil, has to be detoxicated.

The existence in plants of the N-heterocyclic β-substituted alanine lupinic acid (XIV) which is, in effect, the cytokinin zeatin with an alanine residue attached to N[9] (Duke *et al.*, 1978) is another argument in favour of alanine transfer being a plant detoxification mechanism for N-heterocycles. As might be expected if this hypothesis is correct, in contrast to zeatin itself, lupinic acid has only very weak cytokinin activity (Parker *et al.*, 1978), and it has been shown to be formed by the general alanine-transfer mechanism described earlier in this Chap-

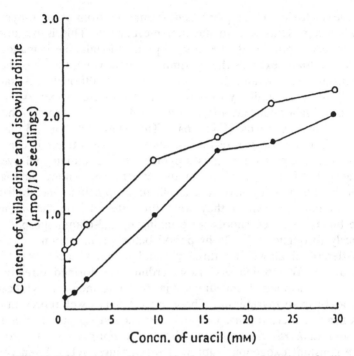

Fig. 17. Effect of uracil concentration on the production of willardiine and isowillardiine by pea seedlings. Seeds were allowed to imbibe at zero time in solutions of uracil of specified concentration. Seedlings were extracted at 136 h and their willardiine and isowillardiine contents determined (Ashworth, Brown & Roberts, 1972; reproduced by kind permission of the *Biochemical Journal*).

ter, that is to say from zeatin and O-acetylserine (Fig. 18) in a pyridoxal 5-phosphate-dependent enzymic reaction (Murakoshi *et al.*, 1977).

Consideration of the N-heterocyclic β-substituted alanines, their biosynthesis and possible functions, serves to highlight the many major gaps in current knowledge of this fundamentally important interface between primary and secondary metabolism. In particular, it points to the existence of a number of plant enzymes of N-heterocyclic synthesis that have high potential for biotechnological exploitation, especially in industrial biotransformation processes (Brown, 1993). There is, undoubtedly, increasing need to know more of the regulation of the interconnections between primary and secondary metabolic pathways and control of the production of plant secondary products, many of which are compounds of high commercial value (Brown, 1985).

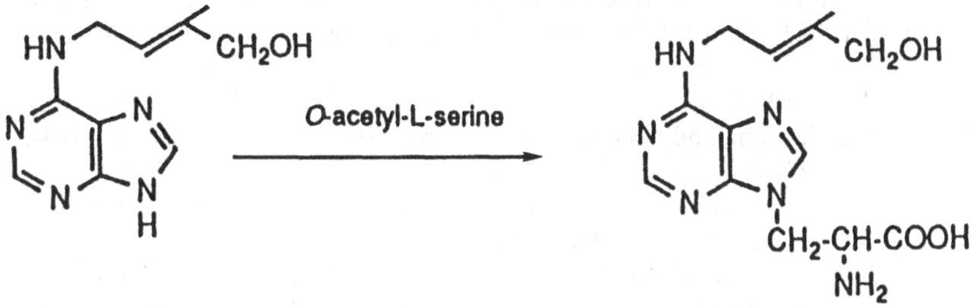

Fig. 18. Enzymic formation of lupinic acid by transfer of an alanine residue from O-acetylserine to zeatin (Murakoshi *et al.*, 1977).

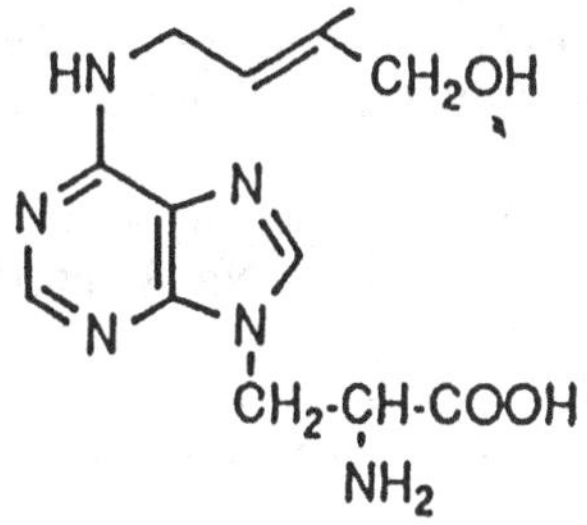

XIV

References

Ahmmad, M.A.S., Maskall, C.S. & Brown, E.G. (1984). Partial purification and properties of willardiine and isowillardiine synthase activity from Pisum sativum. Phytochemistry, **23**, 265–70.

Al-Baldawi, N.F. & Brown, E.G. (1983). Metabolism of [6-^{14}C]orotate by shoots of *Pisum sativum, Phaseolus vulgaris* and *Lathyrus tingitanus. Phytochemistry*, **22**, 1925–8.

Al-Baldawi, N.F.H. & Brown, E.G. (1985). Metabolism of the amino acid β-pyrazol-l-ylalanine and its parent base pyrazole. *Biochemical Pharmacology*, **34**, 1273–8.

Ashworth, T.S., Brown, E.G. & Roberts, F.M. (1972). Biosynthesis of willardiine and isowillardiine in germinating pea seeds and seedlings. *Biochemical Journal*, **129**, 897–905.

Bell, E.A. (1963). A new amino-acid, γ-hydroxyhomoarginine in *Lathyrus*. *Nature*, **199**, 70–1.

Bell, E.A. & Przbylska, J. (1965). The origin and site of synthesis of the pyrimidine ring in the amino acid lathyrine. *Biochemical Journal*, **94**, 35P.

Brown, E.G. (1970). Aspects of the biosynthesis of amino acids in higher plants. *Phytochemistry*, **9**, 122–3.

Brown, E.G. (1975). Pyrimidine derivatives in higher plants. *Phytochemistry*, **14**, 856.

Brown, E.G. (1985). Commercial significance of plants: history and pointers to the future. In *Plant Products and the New Technology*, ed. K.W. Fuller & J.R. Gallon, pp. 1–10. Oxford: Oxford University Press.

Brown, E.G. (1993). Enzymes of N-heterocyclic ring synthesis. In *Methods in Plant Biochemistry*, ed. P.J.Lea. Vol.9, pp.153–81. London: Academic Press.

Brown, E.G. & Al-Baldawi, N.F. (1977). Biosynthesis of the pyrimidinyl amino acid lathyrine by *Lathyrus tingitanus* L. *Biochemical Journal*, **164**, 589–94.

Brown, E.G. & Diffin, F.M. (1990). Biosynthesis and metabolism of pyrazole by *Cucumis sativus*: enzymic cyclization and dehydrogenation of 1,3-diaminopropane. *Phytochemistry*, **29**, 469–78.

Brown, E.G., Flayeh, K.A.M. & Gallon, J.R. (1982). The biosynthetic origin of the pyrazole moiety of β-pyrazol-l-ylalanine. *Phytochemistry*, **21**, 863–7.

Brown, E.G. & Mohamad, J. (1990). Biosynthesis of lathyrine; a novel synthase activity. *Phytochemistry*, **29**, 3117–21.

Cederbaum, A.I. & Rubin, E. (1974). Effects of pyrazole, 4-bromopyrazole and 4-methylpyrazole on mitochondrial function. *Biochemical Pharmacology*, **23**, 203–13.

Duke, C.C., MacLeod, J.K., Summons, R.E., Letham, D.S. & Parker, C.W. (1978). The structure and synthesis of cytokinin metabolites. II. Lupinic acid and O-β-D-glycopyranosylzeatin from *Lupinus angustifolius*. *Australian Journal of Chemistry*, **31**, 1291–301.

Dunnill, P.M. & Fowden, L. (1963). The biosynthesis of β-pyrazol-l-ylalanine. *Journal of Experimental Botany*, **14**, 237–48.

Flayeh, K.A.M., Najafi, S.I., Al-Delymi, A.M. & Hajar, M.A. (1984). 1,3-Diaminopropane and spermidine ln *Cucumis sativus* (cucumber). *Phytochemistry*, **23**, 989–90.

Fowden, L. (1967). Aspects of amino acid metabolism in plants. *Annual Review of Plant Physiology,* **18**, 85–106.

Frisch, D.M., Dunnill, P.M., Smith, A. & Fowden, L. (1967). The specificity of amino acid biosynthesis in the Cucurbitaceae. *Phytochemistry,* **6**, 921–31.

Giovanelli, J. & Mudd, S.H. (1967). Synthesis of homocysteine and cysteine by enzymic extracts of spinach. *Biochemical and Biophysical Research Communications,* **27**, 150–6.

Hadwiger, L.A., Floss, H.G., Stoker, 3.R. & Conn, E.E. (1965). Biosynthesis of orcylalanine. *Phytochemistry,* **4**, 825–30.

Hyde, C.C., Ahmed, S.A., Padlan, E.A., Miles, E.W. & Davies, D.R. (1988). Three-dimensional structure of the tryptophan synthase $\alpha_2\beta_2$ multienzyme complex from *Salmonella typhimurium. Journal of Biological Chemistry,* **263**, 17857–71.

Ikegami, F., Kaneko, M., Kamiyama, H. & Murakoshi, I. (1988). Purification and characterization of cysteine synthases from *Citrullus vulgaris. Phytochemistry,* **27**, 697–701.

Ikegami, F., Kaneko, M., Lambein, F., Kuo, Y-H. & Murakoshi, I. (1987). Difference between uracilalanine synthases and cysteine synthases in *Pisum sativum. Phytochemistry,* **26**, 2699–704.

Li, T.K. & Theorell, H. (1969). Human liver alcohol dehydrogenase: inhibition by pyrazole and pyrazole analogues. *Acta Chemica Scandinavica,* **23**, 892–902.

Lieber, C.S., Rubin, E., Decarli, L.M., Misra, P. & Gang, (1970). Effects of pyrazole on hepatic function and structure. *Laboratory Investigation,* **22**, 615–21.

MacDonald, E., Marselos, M. & Nousiainen, U. (1975). Central and peripheral catechol amine levels after pyrazole treatment. *Acta Pharmacologica et Toxicologica,* **37**, 106–12.

Marselos, M., Törrönen, P., Alakuijala, P. & MacDonald, E. (1977). Hepatic hydroxylation and glucuronidation in the rat after subacute pyrazole treatment. *Toxicology,* **8**, 251–61.

Metzler, D.E., Ikawa, M. & Snell, E.E. (1953). A general mechanism for vitamin B_6-catalysed reactions. *Journal of the American Chemical Society,* **76**, 648–52.

Murakoshi, I., Ikegami, F., Ookawa, N., Ariki, T., Haginiwa, J., Kuo, Y-H & Lambein, F. (1978). Biosynthesis of the uracilylalanines willardiine and isowillardiine in higher plants. *Phytochemistry,* **17**, 1571–6.

Murakoshi, I., Ikegami, F., Ookawa, N., Haginiwa, J. & Letham, D.S. (1977). Enzymatic synthesis of lupinic acid, a novel metabolite of zeatin in higher plants. *Chemical and Pharmaceutical Bulletin,* **25**, 520–2.

Murakoshi, I., Kuramoto, H., Haginiwa, J. & Fowden, L. (1972). The enzymic synthesis of β-substituted alanines. *Phytochemistry,* **11**, 177–82.

Noe, F.F. & Fowden, L. (1960). β-Pyrazol-l-ylalanine, an amino acid from water-melon seeds *(Citrullus vulgaris)*. *Biochemical Journal*, **77**, 543–7.

Ngo, T.T. & Shargool, P.D. (1974). Enzymic synthesis of L-cysteine in higher plant tissues. *Canadian Journal of Biochemistry*, **52**, 435–40.

Nishimura, S., Taga, Y., Kuchino, Y. & Ohashi, Z. (1974). Enzymatic synthesis of 3-(3-amino-3-carboxypropyl)uridine in *Escherichia coli* phenylalanine transfer RNA: transfer of the 3-amino-3-carboxypropyl group from S-adenosylmethionine. *Biochemical and Biophysical Research Communications*, **57**, 702–8.

Noguchi, M., Koiwai, A. & Tamaki, E. (1966). Studies on nitrogen metabolism in tobacco plants. VIII. Δ-pyrroline-5-carboxylate reductase from tobacco leaves. *Agricultural Biological Chemistry*, **30**, 452–6.

Nowacki, E. & Nowacka, D. (1963). Biosynthesis of tingitanine. A free amino acid from Tangier pea. *Bulletin de 1 'Académie Polonaise des Sciences. Série des Sciences Biologiques*, **11**, 361–3.

Ohashi, Z., Maeda, M., McCloskey, J.A. & Nishimura, S. (1974). 3-(3-Amino-3-carboxypropyl)uridine, a novel modified nucleoside isolated from *Escherichia coli* phenylalanine transfer ribonucleic acid. *Biochemistry*, **13**, 2620–5.

Parker, C.W., Letham, D.S., Gollnow, B.I., Summons, R.E., Duke, C.C. & MacLeod, J.K. (1978). Regulators of cell division in plant tissues. XXV. Metabolism of zeatin by lupin seedlings. *Planta*, **142**, 239–51.

Takeshita, M., Nishizuka, Y. & Hayaishi, O. (1963). Studies on β-(pyrazolyl-N)-L-alanine. *Journal of Biological Chemistry*, **238**, 660–5.

Theorell, H. & Yonetani, T. (1963). Liver alcohol dehydrogenase-diphosphopyridine nucleotide (DPN)-pyrazole complex; a model of an intermediate in the enzyme reaction. *Biochemische Zeitschrift*, **338**, 537–53.

Theorell, H., Yonetani, T. & Sjoberg, B. (1969). Effects of some heterocyclic compounds on the enzymic activity of liver alcohol dehydrogenase. *Acta Chemica Scandinavica*, **23**, 255–60.

Tiwari, H.P., Penrose, W.R. & Spenser, I.D. (1967). Biosynthesis of mimosine: incorporation of serine and of α-aminoadipic acid. *Phytochemistry*, **9**, 122–3.

Yon, R.J. (1971a). End-product inhibition of aspartate carbamoyl transferase from wheat germ. *Biochemical Journal*, **121**, 18P–19P.

Yon, R.J. (1971b). Allosteric properties of wheat germ aspartate transcarbamoylase. *Biochemical Journal*, **124**, 10P–11P.

J.P.F. D'MELLO

Toxicity of non-protein amino acids from plants

Introduction

Over 200 non-protein amino acids occur naturally in plants (Rosenthal, 1982). With a few exceptions, these amino acids exist in unconjugated forms and many are associated with toxic properties (D'Mello, 1991). Legumes contain higher concentrations and a more diverse range of non-protein amino acids than any other plant species, and the seed is generally the most concentrated source of these substances.

The toxic non-protein amino acids are a distinguishing feature of many tropical legumes (D'Mello, 1992), contributing significantly to the noxious effects of a number of grain and forage legumes, including *Canavalia ensiformis, Indigofera spicata, Leucaena leucocephala* and at least three *Lathyrus* species. However, there is evidence of considerable variation in the concentrations of non-protein amino acids in different species of the same genus (Aylward *et al.*, 1987). Toxic non-protein amino acids also occur in non-leguminous plants, both tropical (e.g. *Blighia sapida; Cycas circinalis)* and temperate (e.g. *Brassica* species).

The toxicity of non-protein amino acids has been observed in insects, laboratory and farm animals and in humans, but there are striking differences among animal species in their sensitivity to these compounds. Factors such as diet, duration of feeding and geographical differences in microbial ecology in ruminants may also modulate the response to these amino acids.

The role of the non-protein amino acids in plants remains a matter of some debate. However, there is mounting evidence that these compounds form part of the chemical defence system against predation and disease (Rosenthal & Bell, 1979).

The classification of the non-protein amino acids is generally based on their structural relationships with the nutritionally important amino acids and their metabolites and on their physiological properties. Thus some amino acids are structural analogues of urea cycle intermediates such as arginine and ornithine. Other non-protein amino acids are grouped together on the basis of their structural analogy with the aromatic amino acids or with those containing sulphur. Another group of non-protein amino acids are recognised for the neurotoxic properties.

Distribution and toxicity

The distribution and toxicity of the major non-protein amino acids are summarized in Table 1. Canavanine, a structural analogue of arginine, is the most ubiquitous of all these amino acids, occurring in the seed of tree legumes *(Dioclea megacarpa)* as well as forage legumes *(Indigofera spiccata)*. The seed of the latter legume also contains another arginine analogue in the form of indospicine but there are many accessions of this legume which are devoid of indospicine (Aylward *et al.*, 1987). A third analogue of arginine, homoarginine occurs in *Lathyrus cicera*.

Complex interactions and antagonisms exist among these analogues in their relationship with arginine. The metabolism of canavanine illustrates the complexity of these interactions. Its immediate degradation product, canaline, is an analogue of ornithine, the breakdown product of arginine. Indeed, the structural similarities are such that enzyme systems in diverse organisms are unable to distinguish between respective analogues. Thus arginase uses both arginine and canavanine as substrates as does arginyl-tRNA synthetase (Allende & Allende, 1964). Insects sensitive to canavanine possess an arginyl-tRNA synthetase which cannot discriminate between the two analogues with the consequence that canavanine is incorporated into tissue proteins causing pupal malformations and reduced survival of these insects, (Rosenthal, 1977). Not unexpectedly, canavanine acts as a potent chemical barrier to insect predators, particularly when present in high concentrations, as in the seed of *Dioclea megacarpa* (Table 1). Among higher animals, avian species are particularly sensitive to canavanine toxicity since they rely totally upon a dietary supply of arginine. This toxicity is readily demonstrated by feeding diets containing autoclaved *Canavalia ensiformis*. The heat-treatment is necessary to inactivate concanavalin A, a powerful lectin. Young chicks fed such diets show rapid and severe reductions in growth and efficiency of nitrogen utilization with a dietary canavanine concentration of only 3.7 g/kg. These effects are largely reproduced on feeding similar quantities of canavanine. Not unexpectedly, the toxicity of canavanine in young chicks is influenced by dietary factors, being exacerbated by the presence of a small excess of lysine whereas arginine supplementation partially alleviates the adverse effects of canavanine (D'Mello, 1991).

Table 1. *Distribution and toxicity of some non-protein acids*

Amino acid	Legume species	Concentration (g/kg dry weight)	Toxic effects
Arginine analogues			Pupal malformation
Canavannine	*Canavalia*		and reduced survival
	ensiformis	51 (seed)	of insects; reduced
	Gliricidia sepium	40 (seed)	food intake, and N
	Robinia	98 (seed)	utilization;
	pseudoacacia		dysfunction of
	Indigofera spicata	9 (seed)	immune system;
	Dioclea megacarpa	127 (seed)	alopecia
Indospicine	*Indigofera spicata*	20 (seed)	Teratogenic effects; liver damage
Homoarginine	*Lathyrus cicera*	12 (seed)	Reduced growth and food intake
Aromatic			Loss of wool;
Mimosine	*Leucaena*	145 (seed)	teratogenic effects;
	Leucocephala	25 (leaf)	organ damage; death
Analogues of sulphur amino acids			
Se-methylseleno-cysteine	*Astragalus*	?	'Blind staggers'; death
Selenocystathionine			
Selenomethionine			
S-methylcysteine sulphoxide	*Brassica* species	40–60 (forage)	Haemolytic anaemia; organ damage; death
Neurotoxic			
β-cyanoalanine	*Vicia sativa*	1.5 (seed)	Convulsions; tetanic spasms; death.
β-(N) oxalyl amino alanine	*Lathyrus sativus*	25 (seed)	Neurolathyrism.
α,γ-diaminobutyric acid	*Lathyrus latifolius*	16 (seed)	Hyperirritability; tremors; death.
β-N-methyl amino acid	*Cycas circinalis*	?	Guam dementia
Miscellaneous			Vomiting sickness
Hypglycin A	*Blighia sapida*	1.1 (unripened fruit)	causing convulsions and death

Of greater economic significance is the toxicity of the aromatic amino acid, mimosine (Table 1) which occurs in the leaves and seeds of *Leucaena leucocephala,* an ubiquitous tropical legume capable of yielding prolific quantities of palatable forage for ruminant animals. The deleterious properties of mimosine include disruption of reproductive processes, loss of hair and wool, organ damage and death (Reis, Tunks & Chapman, 1975; Reis, Tunks & Hegarty, 1975). The manifestation of toxicity in ruminants is dependent upon the rate and extent of microbial degradation of mimosine to 3-hydroxy-4(1H)-pyridone (3,4-DHP) and to its isomer 2,3-DHP (Jones, 1985). In addition, some rumen microbes are capable of degrading the two forms of DHP to, as yet unidentified, non-toxic residues. Ruminant animals in Australia and the USA lack the requisite bacteria for DHP detoxification and are, therefore, susceptible to the goitrogenic effects of these by-products (Jones, 1985; Hammond et al., 1989). However, in certain other regions where *Leucaena* is indigenous (Central America) or is naturalized (Hawaii and Indonesia), ruminants possess the full complement of bacteria that are required for DHP detoxification, which explains the absence of adverse effects associated with *Leucaena* in these countries.

Striking analogues of the sulphur-containing amino acids exist in various plant species (Table 1). In *Astragalus* species, the sulphur atom is replaced by selenium. These seleno-amino acids are associated with debilitating effects in ruminants grazing *Astragalus* and other Se-rich plants (Stadtman, 1974). In forage and root *Brassica* crops, the non-protein amino acid, S-methylcysteine sulphoxide (SMCO) occurs in sufficient concentrations to precipitate a severe haemolytic anaemia in ruminants grazing these crops. The toxicity of SMCO arises after its conversion by rumen microbes to dimethyl disulphide (Smith, 1980). External manifestations of this disorder include loss of appetite and reduced milk production. Internally, the appearance of refractile, stainable granules within the erythrocytes and a decline in blood haemoglobin concentrations may be seen. Extensive organ damage normally accompanies this condition, the liver becoming swollen, pale and necrotic. Spontaneous but incomplete recovery may occur in survivors continuing to graze the *Brassica* crop but further fluctuations in blood haemoglobin concentrations may ensue.

A number of neurotoxic amino acids occur in leguminous and other species of plants. These include the neurolathyrogens, β-cyanoalanine, β-(N) oxalyl amino alanine (BOAA) and α,γ-diaminobutyric acid (Table 1). These amino acids are believed to be responsible for the condition, neurolathyrism, in humans. The disorder is characterized by muscular rigidity, weakness, paralysis of leg muscles and in some cases, death (Padmanaban, 1980). Another amino acid, β-N-methyl amino alanine, structurally analogous to BOAA, also evokes neurological disorders in humans consuming seeds of the non-legume, *Cycas circinalis* (Spencer *et al.*, 1987).

In Jamaica consumption of the unripened fruit of *Blighia sapida* which contains more hypoglycin A than the ripened fruit (Hassal & Reyle, 1955), causes a serious disorder in humans known as 'vomiting sickness'. In addition to the emetic effects, this amino acid also causes convulsions, coma and death arising from a marked hypoglycaemia in affected subjects.

Mode of action

There is considerable evidence to indicate that the toxic action of individual non-protein amino acids is mediated via a diverse array of mechanisms (D'Mello, 1991). In the case of canavanine, the multimodal action may involve increased renal and hepatic arginase activities; inhibition of key enzymes (e.g. ornithine decarboxylase, D'Mello, 1993) dependent upon pyridoxal phosphate through the synthesis of canaline; competition with arginine and/or lysine for intestinal transport; inhibition of transamidinase activity leading to reduced creatine synthesis; and the synthesis of tissue proteins containing canavanine instead of arginine (Fig. 1). The synthesis of anomalous canavanyl proteins is the primary means whereby canavanine-sensitive insects such as the tobacco hornworm (*Manduca sexta*) succumb to the toxic action of this amino acid. In higher animals, however, there is evidence that other mechanisms are more important. For example, in avian species enhanced arginine degradation would be critical owing to the indispensability of this amino acid for birds. A more recent study (D'Mello, 1993) showed that liver ornithine decarboxylase activity in chicks fed seeds of *Canavalia ensiformis* declined to only 19% of that observed in control animals causing a marked accumulation of ornithine in serum. This may be another important mechanism of canavanine toxicity in higher animals.

The action of other toxic non-protein amino acids also involves diverse mechanisms and these are reviewed by D'Mello (1991).

Detoxification

The ability to detoxify non-protein amino acids appears to be an innate feature of many organisms. Microbial detoxification of mimosine occurs in ruminants in regions where *Leucaena* is indigenous or is naturalized. In other parts of the tropics (e.g. Australia and USA) ruminants lack the requisite bacteria required for DHP detoxification. However, the

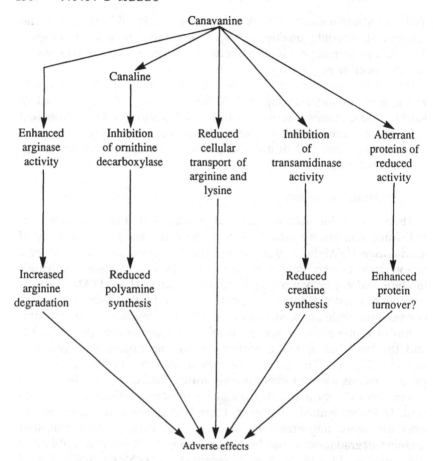

Fig. 1. Diverse mechanisms underlying the toxicity of canavanine (D'Mello, 1991; reproduced by permission of the Royal Society of Chemistry).

transfer of DHP-degrading bacteria to cattle in Australia and the USA has been accomplished. Dosed animals grazing *Leucaena* showed markedly improved growth and serum thyroxine (T4) concentrations than untreated Leucaena-fed steers (Quirk *et al.*, 1988).

In animals, complete detoxification only occurs if they possess the versatile metabolic machinery required to counteract the multi-modal action of non-protein amino acids. Thus the tropical bruchid beetle, *Caryedes brasiliensis* avoids canavanine toxicity by virtue of enzyme systems capable of distinguishing between canavanine and arginine during protein synthesis (Rosenthal, Dahlman & Janzen, 1977). In addition, the insect degrades canavanine to canaline and urea. Canaline is detoxified to homoserine, while the ammonia produced in this reaction together with that arising from the action of urease on urea is incorporated into glutamine.

In mammals and birds, the biochemical versatility for amino acid detoxification is considerably reduced. However, in farm animals, considerable mitigation of adverse effects is possible by controlled feeding of forages containing toxic non-protein amino acids. Thus, current recommendations for averting SMCO toxicity in cattle and sheep involve control of *Brassica* forage intake, particularly in the winter when SMCO concentrations are high (Smith, 1980). Grazing management is also an effective means of controlling *Leucaena* toxicity (Jones, 1985).

Detoxification of crop plants containing non-protein amino acids has been the subject of several investigations. For example, there is some interest in the use of plant breeding techniques to reduce or eliminate toxicity associated with these amino acids. However, it should be recognized that these compounds may confer insect and fungal resistance to the plant and the removal of such defence substances may result in agronomic disadvantages. Since the toxic non-protein amino acids occur in unconjugated forms, their extraction from seeds with solvents or aqueous solutions is possible. Thus D'Mello and Walker (1991) showed that seeds of *Canavalia ensiformis* extracted with dilute $KHCO_3$ solution at 80 °C contain virtually no canavanine after 48 hours of treatment. Once autoclaved, these seeds may be incorporated into diets and fed to broiler chicks without detrimental effects. Extraction of *Lathyrus* seeds with hot water also results in the removal of BOAA (Padmanaban, 1980). The relative ease with which amino acids may be extracted from intact beans is noteworthy and may form the basis for the detoxification of other plants containing these compounds.

Conclusions

The toxic non-protein amino acids are a distinctive feature of many tropical legumes. In general, these plants contain higher concentrations and a more diverse array of these substances than any other plant species. However, in temperate countries, S-methylcysteine sulphoxide (SMCO) contributes significantly to the toxicity of *Brassica* forage. The toxic effects of non-protein amino acids in farm animals are wide-ranging. Thus mimosine, present in *Leucaena leucocephala*, induces loss of wool, organ damage and death, while SMCO causes haemolytic anaemia in ruminants fed mainly or solely on *Brassica* forage. The lathyrogenic amino acids present in seeds of certain *Vicia* and *Lathyrus* species and β-N-methyl amino alanine present in seeds of *Cycas circinalis* are associated with potent neurotoxic properties in humans. In ruminants, the presence of symbiotic microorganisms within the rumen sometimes confers protection to the host animal by eliciting degradation of the toxic amino acid to innocuous compounds. However, in the case of SMCO, ruminal metabolism precipitates toxicity through the synthesis of dimethyl disulphide, a highly reactive product.

Considerable alleviation of adverse effects in ruminants may be achieved by controlled feeding or grazing of forages containing toxic non-protein amino acids. Thus, Van Eys *et al.* (1986) were able not only to avoid toxicity but also to increase growth rates of goats by feeding limited quantities of *Leucaena* forage as a supplement to napier grass.

A critical assessment of the available evidence indicates that the toxic action of individual non-protein amino acids is mediated via a diverse array of mechanisms ranging from inhibition of key enzymes to competition with essential amino acids for transport and protein synthesis.

References

Allende, C.C. Allende, J.E. (1964). Purification and substrate specificity of arginylribonucleic acid synthetase from rat liver. *Journal of Biological Chemistry*, **239**, 1102–6.

Aylward, J.H., Court, R.D., Haydock, K.P., Strickland, R.W. & Hegarty, M.P. (1987). *Indigofera* species with agronomic potential in the tropics. Rat toxicity studies. *Australian Journal of Agricultural Research*, **38**, 177–86 .

D'Mello, J.P.F. (1991). Toxic amino acids. In *Toxic Substances in Crop Plants,* ed. J.P.F. D'Mello, C.M. Duffus and J.H. Duffus, pp. 21–48. Cambridge: The Royal Society of Chemistry.

D'Mello, J.P.F. (1992). Chemical constraints to the use of tropical legumes in animal nutrition. *Animal Feed Science and Technology*, **38**, 237–61.

D'Mello, J.P.F. (1993). Non-protein amino acids in *Canavalia ensiformis* and hepatic ornithine decarboxylase. *Amino Acids*, **4** (212–13).

D'Mello, J.P.F. & Walker, A.G. (1991). Detoxification of jack beans *(Canavalia ensiformis):* studies with young chicks. *Animal Feed Science and Technology*, **33**, 117–27.

Hammond, A.C., Allison, M.J., Williams, M.J., Prine, G.M. & Bates, D.B. (1989). Prevention of *Leucaena* toxicosis of cattle in Florida by ruminal inoculation with 3-hydroxy- 4-(lH)pyridone-degrading bacteria. *American Journal of Veterinary Research,* **50**, 2176–80.

Hassal, C.H. & Reyle, K. (1955). Hypoglycin A and B, two biologically active polypeptides from *Blighia sapida. Biochemical Journal,* **60**, 334–9.

Jones, R.J. (1985). *Leucaena* toxicity and the ruminal degradation of mimosine. In *Plant Toxicology,* ed. A.A. Seawright, M.P. Hegarty, L.F. James and R.F. Keeler, pp. 111–19. Yeerongpilly: Queensland Poisonous Plants Committee.

Padmanaban, G. (1980). Lathyrogens. In *Toxic Constituents of Plant Foodstuffs,* ed. I.E. Liener, pp. 239–63. New York: Academic Press.

Quirk, M.F., Bushell, J.J., Jones, R.J., Megarrity, R.G. & Butler, K.L. (1988). Liveweight gains on *Leucaena* and native grass pastures after dosing cattle with rumen bacteria capable of degrading DHP, a ruminal metabolite from *Leucaena. Journal of Agricultural Science,* **111**, 165–70.

Reis, P.J., Tunks, D.A. & Chapman, R.E. (1975). Effects of mimosine, a potential chemical defleecing agent, on wool growth and the skin of sheep. *Australian Journal of Biological Sciences,* **28**, 69–84.

Reis, P.J., Tunks, D.A. & Hegarty, M.P. (1975). Fate of mimosine administered orally to sheep and its effectiveness as a defleecing agent. *Australian Journal of Biological Sciences,* **28**, 495–501.

Rosenthal, G.A. (1977). The biological effects and mode of action of L-canavanine, a structural analogue of L-arginine. *The Quarterly Review of Biology,* **52**, 155–78.

Rosenthal, G.A. (1982). *Plant Nonprotein Amino Acids and Imino Acids.* 273 pp. New York: Academic Press.

Rosenthal, G.A. & Bell, E.A. (1979). Naturally occurring, toxic nonprotein amino acids. In *Herbivores: Their Interaction with Secondary Plant Metabolites,* ed. G.A. Rosenthal and D.H.Janzen, pp. 353–85. New York: Academic Press.

Rosenthal, G.A, Dahlman, D.L. & Janzen, D.H. (1977). Degradation and detoxification of canavanine by a specialized seed predator. *Science,* **196**, 658–60.

Smith, R.H. (1980). Kale poisoning: the Brassica anaemic factor. *The Veterinary Record,* **107**, 1215.

Spencer, R.S., Nunn, P.B., Hugon, J., Ludolph, A.C., Ross, S.M., Roy, D.N. & Robertson, R.C. (1987) . Guam amyotrophic lateral sclerosis – Parkinson-dementia linked to a plant excitant neurotoxin. *Science,* **237**, 517–22.

Stadtman, T.C. (1974). Selenium biochemistry. *Science,* **183**, 915–22.

Van Eys, J.E., Mathius, I.W., Pongsapan, P. and Johnson, W.L. (1986). Foliage of the tree legumes *Gliricidia, Leucaena* and *Sesbania* as supplement to Napier grass diets for growing goats. *Journal of Agricultural Science,* **107**, 227–33.

HEINZ RENNENBERG

Processes involved in glutathione metabolism

Introduction

A major part of our present knowledge on the processes involved in glutathione metabolism originates from research performed by animal and human biochemists and physiologists up to the late 1970s (Meister, 1981; Meister & Anderson, 1983). Until that time interest of plant biochemists and physiologists in research on glutathione was rather low. Apparently, it was assumed that experiments on glutathione metabolism performed with animal tissues had provided answers that are also valid for plants. This situation changed entirely about a decade ago, when it became obvious that glutathione is an important factor in stress physiology of plants (Smith, Polle & Rennenberg, 1990; Rennenberg & Brunold, 1994). Glutathione was found to be an antioxidant, both as a constituent of the chemical defence system and as a substrate of the enzymatic defense system in the cytoplasm and the chloroplasts of the cells (Polle & Rennenberg, 1993, 1994). Glutathione was found to counteract heat, cold and drought stress (Smith *et al.*, 1990). Glutathione was identified as a substrate for the conjugation of xenobiotics (Rennenberg & Lamoureux, 1990) and for the synthesis of phytochelatins, i.e. poly(γ-glutamylcysteinyl)-glycines involved in metal homeostasis and metal detoxification (Rennenberg & Brunold, 1994). Glutathione was found to be involved in signal transduction in plant-pathogen interactions (Rennenberg & Brunold, 1994). Glutathione was found to be a storage form of reduced sulphur, compensating stress from excess sulphur in plants (Rennenberg, 1984; De Kok, 1990). Today, the action of glutathione in stress physiology of plants belongs to the most active areas of research in plant sciences.

It is evident from research performed on glutathione in stress physiology that the processes involved in glutathione metabolism in plants are different from those in animals and human beings in many respects. In addition, other γ-glutamylcysteinyl-tripeptides, like homoglutathione (γ-glutamylcysteinyl-β-alanine) or hydroxymethylglutathione (γ-glut-amylcysteinylserine), can be present in plant cells either together with, or without glutathione (Bergmann & Rennenberg, 1993). Still, our present knowledge on basic biochemical and physiological processes of the metabolism of glutathione and its homologues in plants is far from being sufficient to provide the information required for a proper interpretation of data obtained in stress physiological studies. The aim of this review is to summarize today's understanding of the processes involved in glutathione synthesis and degradation, and its distribution within the plant. Special emphasis will be given to open questions that are of particular importance in stress physiology.

Synthesis of glutathione

Recent experiments with several species indicate that plant cells contain both γ-glutamylcysteine synthetase (γ-GCS) and glutathione synthetase (GSH-S) activity (Bergmann & Rennenberg, 1993). From this finding it is concluded that the synthesis of glutathione in plant cells proceeds in the same sequence of reactions previously described for animal cells (Meister & Anderson, 1983). In an initial reaction γ-glutamylcysteine is synthesized from glutamate and cysteine (I). This reaction is dependent on ATP; it is catalyzed by a γ-glutamylcysteine synthetase (EC 6.3.2.2). Corresponding to the relatively high concentrations of gluta-mate and the relatively low concentrations of cysteine usually found in plant cells, low in vitro affinities for glutamate and high in vitro affinities for cysteine were found for the enzyme (Bergmann & Rennen-berg, 1993).

$$glu + cys + ATP \xrightarrow{\text{γ-GCS}} \text{γ-glu–cys} + ADP + Pi \qquad (I)$$

$$\text{γ-glu–cys} + gly + ATP \xrightarrow{\text{GSH-S}} \text{γ-glu–cys–gly} + ADP + Pi \qquad (II)$$

In a second reaction glycine is added to the C-terminal site of γ-glutamylcysteine to yield glutathione (II). This reaction is also dependent on ATP; it is catalyzed by a glutathione synthetase (EC 6.3.2.3). Although cellular γ-glutamylcysteine concentrations are usually much lower than the cellular concentrations of glycine in plants, in vitro affinities of the enzyme for the two substrates were similar. Therefore, glutathione synthesis may be limited in vivo by the availability of γ-glutamylcysteine (Bergmann & Rennenberg, 1993). In the presence of excess reduced sulphur in the dark, however, glycine may become a limiting factor of glutathione synthesis (Buwalda et al., 1988, 1990).

This effect may be explained by a lack of photorespiratory glycine production at elevated cysteine synthesis. Glutathione synthetase seems to be specific for glycine and does not use β-alanine as a substrate. Apparently, homoglutathione is synthesized by a β-alanine specific homoglutathione synthetase (Bergmann & Rennenberg, 1993). Whether hydroxymethylglutathione is synthesized by a serine specific enzyme, or by processing of glutathione remains to be elucidated.

Both enzymes of glutathione synthesis, γ-glutamylcysteine synthetase and glutathione synthetase, were found to be localized in the chloroplast and in the cytosol in similar activities (Bergmann & Rennenberg, 1993). The actual distribution of glutathione synthesis between the chloroplasts and the cytosol, however, has not been studied. Depending on the source of cysteine this distribution may change during plant development. In mature leaves, e.g., cysteine is predominantly synthesized in the chloroplasts (Giovanelli, 1990) suggesting a predominant synthesis of glutathione in this organelle as well; when cysteine is provided by the degradation of storage protein during the early stages of leaf development, glutathione may be synthesized in the cytosol at a high extent. The enzymes of glutathione synthesis have also been found in root tissue (Rüegsegger & Brunold, 1992). γ-Glutamylcysteine synthetase activity was almost equally distributed between the cytosol and the proplastids; but glutathione synthetase activity was primarily localized in the cytosol. At which extent glutathione synthesis in the roots contributes to the overall glutathione synthesis in the plant is presently unknown.

Several pieces of evidence indicate that the first step of glutathione synthesis, i.e. the synthesis of γ-glutamylcysteine, is subject to feedback inhibition by glutathione. *In vitro*, γ-glutamylcysteine synthetase activity is inhibited by physiological GSH concentrations (Bergmann & Rennenberg, 1993). Exposure of plant cells to Cd results in a decline of the glutathione content accompanied by intensive phytochelatin synthesis from glutathione. When the rate of glutathione synthesis is calculated from the amount of phytochelatins produced, a 5- to 8-fold increase is found at exposure to Cd. The glutathione content itself was maintained at a low level under these conditions (Bergmann & Rennenberg, 1993). It can therefore be assumed that a high rate of glutathione synthesis is obtained by a partial release of the feedback inhibition of γ-glutamylcysteine synthetase by GSH. This release may be attributed to a reduced glutathione concentration as a consequence of intensive use of glutathione in phytochelatin synthesis. This regulatory mechanism may be means of plants to rapidly react to enhanced requirements of glutathione. However, data on the fluxes of cysteine into γ-glutamylcysteine, glutathione and phytochelatins in the presence and the absence of Cd to prove this assumption are lacking. Recent experiments in the author's laboratory with transgenic poplar plants indicate that the feedback control of glutathione synthesis may be very strict. Although cytoplasmic glutathione synthetase activity of the leaves was elevated more than 50-fold, cellular glutathione levels remained unchanged (Strohm, unpublished results).

On the other hand, it has been observed by several authors that cellular glutathione levels increase as a consequence of various environmental stresses such as excess sulphur (De Kok, 1990; Smith *et al.*, 1990), high elevation (Polle & Rennenberg, 1994), or exposure to safeners (Hatzios & Hoagland, 1989; Lamoureux & Rusness, 1989). These findings support the idea that glutathione synthesis may not only be controlled by feedback inhibition, but also at the molecular level. In addition, compartmentation of the glutathione produced may be a means to overcome feedback inhibition of γ-glutamylcysteine synthetase by GSH. Although glutathione concentrations have been determined in the chloroplasts of several species (Rennenberg & Lamoureux, 1990), reliable data on the glutathione content of other cellular compartments, esp. the cytoplasm and the vacuole, are lacking. It can therefore not be decided whether pools of glutathione that are not subject to feedback control exist in plant cells.

Glutathione degradation

A definite pathway of glutathione degradation has so far not been established in plants (Bergmann & Rennenberg, 1993). From experiments with green tobacco cells, it is concluded that degradation of glutathione in plant cells is initiated by a carboxypeptidase removing the C-terminal glycine moiety of glutathione (Fig. 1C). The γ-glutamylcysteine produced this way is further degraded to cysteine and glutamate by the successive action of γ-glutamylcyclotransferase and 5-oxo-prolinase. Whereas γ-glutamylcyclotransferase and 5-oxo-prolinase are both present in green tobacco cells, glutathione specific carboxypeptidase activity has not been demonstrated in plants (Rennenberg & Lamoureux, 1990). The first step of this pathway is, however, consistent with the metabolism of conjugates of pesticides with glutathione in plants, where γ-glutamylcysteine derivatives have frequently been observed as initial degradation products (Lamoureux & Rusness, 1989). Still, the operation of this pathway in plant cells was doubted from the following findings: (1) γ-glutamylcysteine is an intermediate of both, the proposed pathway of glutathione degradation and the pathway of

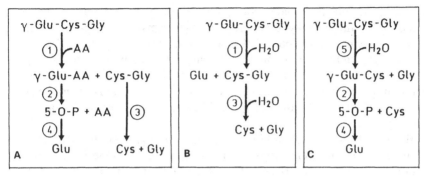

Fig. 1. Pathways of glutathione degradation (modified from Berg-mann & Rennenberg, 1993). (1), γ-glutamyl transpeptidase; (2), γ-glut-amyl cyclotransferase; (3), dipeptidase; (4), 5-oxo-prolinase; (5), carboxypeptidase.

glutathione synthesis; (2) the enzymes of glutathione degradation are entirely, the enzymes of glutathione synthesis are partially, localized in the cytosol; (3) the affinity of γ-glutamylcyclotransferase for γ-gluta-mylcysteine determined *in vitro* was much too low to compete success-fully with GSH synthetase. Therefore, it was assumed that the pathway suggested above would lead to a futile cycle, in which γ-glutamylcysteine produced by glutathione degradation would be used directly for gluta-thione synthesis, thereby preventing degradation of glutathione into the constituent amino acids (Bergmann & Rennenberg, 1993). This assumption is, however, based on the conclusion that glutathione syn-thesis and degradation proceeds simultaneously in the cytosol, because the enzymes of both pathways are present in this cellular compartment. Experimental evidence for this conclusion has so far not been presented. As glycine can become a limiting factor in glutathione synthesis (Buwalda *et al.*, 1988, 1990), the cytosolic concentration of this amino acid may be an important factor controlling whether glutathione syn-thesis or degradation proceeds in this compartment.

Recent studies with plant tissues indicate that another pathway of glutathione degradation can operate in plants cells (Fig. 1B). Evidence for this pathway comes from the findings that (1) cysteinylglycine is an intermediate of glutathione degradation in *Glycine max*, *Hordeum vulgare* and heterotrophic tobacco cells, (2) cysteinylglycine can be produced by the γ-glutamyl transpeptidase present in these species, and (3) cysteinylglycine can be degraded to the constituent amino acids by a dipeptidase found in plant cells (Bergmann & Rennenberg, 1993). From these observations, a pathway of glutathione degradation is assumed in which initially the γ-glutamyl-moiety of glutathione is removed by the hydrolytic activity of a γ-glutamyl transpeptidase. The cysteinylglycine produced in this reaction is then hydrolysed by a dipeptidase yielding cysteine and glycine (Bergmann & Rennenberg, 1993). As cysteinylglycine derivatives of pesticides have recently been found in plant cells (Lamoureux, personal communication), degradation of glutathione conjugates of pesticides may also proceed via this pathway.

Both pathways of glutathione degradation suggested to operate in plant cells are different from the pathway of glutathione degradation established in animal tissues (Fig. 1A; Meister, 1981). To test the operation of the pathways proposed in plant cells, labelling experiments with radioactive glutathione are required in order to find out whether the conflicting results reported are the consequence of species- or tissue-specific differences in glutathione degradation. A first approach with heterotrophic tobacco cells and ^{35}S-glutathione resulted in extensive labelling of the cysteine pool of the cells, but labelling of γ-glutamyl-cysteine or cysteinylglycine could not be detected (Schneider, 1992).

Long-distance transport of glutathione

Glutathione produced in higher plant cells may not only be removed by degradation, but also by long-distance transport in phloem and xylem. First evidence for long-distance transport of glutathione in the phloem came from experiments in which soybean plants were fumigated with ^{35}S-sulphur dioxide (Garsed & Read, 1977a,b). When the honey dew of aphids that collected phloem sap from the fumigated plants was analysed, radio-labelled glutathione was detected. However, it could not be decided whether the labelled glutathione originated directly from the phloem sap, or was synthesized by the aphids from other labelled compounds transported in the phloem. Experiments with tobacco (Rennenberg, Schmitz & Bergmann, 1979), castor bean (Bonas et al., 1982), and cucurbit plants (Rennenberg & Thoene, 1987) showed that glutathione is transported in the phloem from mature leaves to the roots, irrespective of whether the roots or the leaves were supplied with radiolabelled sulphate. Injection of ^{35}S-cysteine into the kernel of developing maize seedlings resulted in phloem transport of radiolabelled glutathione from the kernel to the roots (Rauser, Schupp & Rennenberg, 1991). Phloem transport of glutathione to the new needle generation was observed, when ^{35}S-glutathione was fed to one year-old spruce needles (Schupp et al., 1992). Apparently, glutathione seems to be transported in the phloem in its reduced form (GSH) (Bonas et al.,

1982) that can be maintained by glutathione reductase activity present in the transport tissue (Alosi, Melroy & Park, 1988). The glutathione content in the phloem has only been determined in cucurbits, where millimolar concentrations were found in phloem exudate (Alosi *et al.*, 1988). The rate of glutathione transport in the phloem has been estimated from experiments with maize seedlings (Rauser *et al.*, 1991) and spruce trees (Schupp *et al.*, 1992). Phloem transport of glutathione from the maize kernel to the roots proceeded at an estimated rate (27–40 nmol GSH per g fresh weight per hour) in the same order of magnitude as phloem transport from one year old spruce needles to the new needle generation (2,5–11,3 nmol GSH per g needle fresh weight per hour).

In experiments with tobacco plants, radiolabelled glutathione was not only transported from mature leaves to the roots, but also to the young, developing leaves (Rennenberg *et al.*, 1979). In a similar way, export of glutathione from the kernel of maize seedlings proceeded both to the root and to the shoot (Rauser *et al.*, 1991). From the estimated fluxes of glutathione, the sink strength of the roots and the shoot were similar; however, within the shoot young leaves were preferable sinks as compared to older leaves (Rauser *et al.*, 1991). From these experiments, it could not be decided whether glutathione transport to the young developing leaves proceeds in the phloem or in the xylem. First evidence for a transport of glutathione in the xylem came from studies with spruce trees (Schupp *et al.*, 1992). When ^{35}S-glutathione was fed to a one year old needle, radiolabelled glutathione was initially exported in the phloem and transported to the new needle generation. Along the transport path, an increasing amount of radioactivity moved from the bark to the wood with increasing distance from the needle fed ^{35}S-glutathione. Apparently, glutathione was exchanged from the phloem to the xylem along the transport path. Analysis of xylem sap from spruce twigs confirmed that glutathione is the predominant reduced sulphur compound transported in the xylem of spruce trees (Schupp, 1991). Although the concentrations of glutathione in the xylem sap were low (0.5–2.5μM), the actual amount of glutathione transported can be high due to high transport velocities in the xylem. In a recent set of experiments in our laboratory (Schneider, Schatten & Rennenberg, 1994*b*) radiolabelled glutathione was fed to a one year-old spruce needle on a twig girdled apical to the application site. A radial transfer of radioactivity from the bark to the wood basal to the girdle and a redistribution from the wood to the bark apical to the girdle was observed. Apparently, the girdle imposed apical to the application site of radiolabelled glutathione did not hinder long-distance transport.

These findings indicate that glutathione can readily be exchanged in spruce between phloem and xylem in both directions.

However, bidirectional phloem to xylem exchange seems not to be a general phenomenon in plants. In beech trees cysteine rather than glutathione is the predominant thiol in the xylem sap (Schupp, Glavac & Rennenberg, 1991). When radiolabelled cysteine that is rapidly incorporated into glutathione (Schütz, De Kok & Rennenberg, 1991) was fed to beech leaves, radioactivity was almost exclusively found in the bark basipetal to the application site. Apparently, phloem to xylem exchange of reduced sulphur compounds does not take place in beech trees (Herschbach, personal communication). These results obtained with beech cannot be readily transferred to other deciduous trees. In poplar, high amounts of glutathione were found in the xylem sap (up to 13 μM). The presence of appreciable amounts of glutathione was, however, restricted to the time of catkin development in March. At this time of the year glutathione was the predominant reduced sulphur compound transported in the xylem. During bud break and leaf expansion methionine was present in much higher concentrations (up to 13 mM). Apparently, the presence of glutathione in the xylem sap is under ontogenetic control in poplar (Schneider et al., 1994a).

The significance of glutathione in whole-plant sulphur nutrition

The finding of glutathione transport, both in phloem and in xylem, suggests that glutathione is an important factor in the distribution of reduced sulphur within the plant. Because of the spacial separation of sulphate uptake (roots), sulphate reduction (mature leaves), and the use of reduced sulphur in protein synthesis (developing leaves and roots) such a distribution appears to be an essential component of the plants' sulphur nutrition. Comparable to sucrose being the predominant transport form of carbohydrate in many plants, glutathione seems to be the predominant transport form of reduced sulphur. Several pieces of evidence support this assumption. In several plant species glutathione was transported from mature leaves (Rennenberg et al., 1979; Bonas et al., 1982; Rennenberg & Thoene, 1987) or seeds (Rauser et al., 1991) to the roots; sulphur originating from glutathione was found to be incorporated into root protein. This finding is consistent with the observation of low activities of the enzymes of sulphate reduction in the roots (Brunold, 1990). Apparently, roots are unable to meet their own needs for reduced sulphur and are dependent on reduced sulphur import from other organs. In seedlings, the reduced sulphur stored in the seed is initially used to support the growth of the developing roots (Rauser *et al.*, 1991) despite significant activities of the enzymes of sulphate reduction in the roots (Brunold, 1990). Glutathione is also transported from mature leaves to developing leaves (Rennenberg *et al.*, 1979; Bonas *et al.*, 1982; Rennenberg & Thoene, 1987; Schupp & Rennenberg, 1992) and is incorporated into the protein of young leaves (Schupp *et al.*, 1992). In spruce, low activities of the enzymes of sulphate reduction during major parts of development of the new needle generation explain the requirement for glutathione transport to the young sprouts (Schupp & Rennenberg, 1992). Also, seeds can become a sink of reduced sulphur, dependent on the import from external sources, during their development. Again, glutathione or its homologues may be transported to the developing seeds to meet their requirement for reduced sulphur (Macnicol & Bergmann, 1984).

From the finding of glutathione in both phloem and xylem, in some species it may be assumed that glutathione undergoes an internal cycling inside the plant (Fig. 2), as previously observed for reduced nitrogen compounds like amino acids (Gojon *et al.*, 1986; Touraine, Grignon & Grignon, 1990). Plant internal cycling – with mature leaves feeding into the cycle and developing leaves, seeds and roots acting as sinks – would provide means to fulfil the different requirements of plant organs for reduced sulphur and its changes during growth and development. In those species transporting glutathione in the phloem and cysteine and/or methionine in the xylem, glutathione degradation in the roots may act as an additional site of control. Further experiments are required to test this hypothesis.

Recently, an additional function of glutathione transport has been suggested from experiments with heterotrophic and green tobacco suspension cultures (Rennenberg *et al.*, 1988). In these experiments it was observed that the influx of sulphate was inhibited by glutathione in its reduced form (GSH) in heterotrophic, but not in green tobacco cells. This result was achieved with growing cultures and in short-term transport experiments as well. Recovery experiments indicated that GSH can inhibit *de novo* synthesis of sulphate transport entities that undergo rapid turnover (Rennenberg, Kemper & Thoene, 1989). From these experiments, it was concluded that GSH transported in the phloem from mature leaves to the roots may control sulphate uptake of the plants. Inhibition of sulphate influx and xylem loading of sulphate by glutathione was confirmed in experiments with excised tobacco roots (Herschbach & Rennenberg, 1991). More than 70% inhibition was achieved by exposure to 0.1 mM GSH within one hour. However, GSH may not be a highly specific inhibitor of sulphate influx (Gunz,

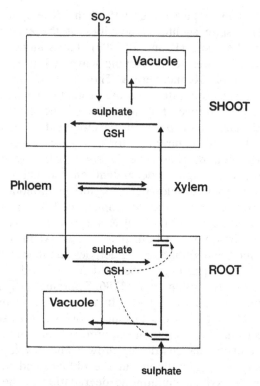

Fig. 2. Plant internal cycling of glutathione (GSH) and sulphate and its interaction with SO₂ and sulphate uptake (modified from: Rennenberg & Polle, 1994) ———> Transport paths of sulphate and GSH; – – – – –> inhibition of sulphate uptake and xylem loading of sulphate by GSH.

Herschbach & Rennenberg, 1992). In the presence of 0.1 mM γ-glutamylcysteine, ophthalmic acid, or oxidized glutathione (GSSG) sulphate influx into excised tobacco roots was also diminished significantly. The percentage inhibition was about half the inhibition found with GSH. Also exposure of excised roots to the sulphur-free amino acids L-alanine, L-aspartate, or L-glutamine in 0.1 mM concentrations resulted in a 27 to 42% reduction of sulphate influx. This observation may be evidence for an inter-pathway control of nitrogen and sulphur metabolism at the level of sulphate influx. Neither the sulphur compounds nor the amino acids mentioned above inhibited the percentage of the sulphate taken up that was loaded into the xylem. It may therefore be concluded that GSH mediated inhibition of sulphate influx into the roots is relatively unspecific, whereas GSH mediated inhibition of xylem loading is a relatively specific process.

When the root system of intact tobacco plants was exposed to 0.1 mM GSH, sulphate influx was reduced by 45% within one hour, but – in contrast to the experiments with excised tobacco roots, the percentage of the sulphate taken up that was loaded into the xylem was not significantly affected (Herschbach, 1992). It may be argued that the external exposure of an intact plant to glutathione is not an appropriate experimental system to study the effect of glutathione transported from mature leaves to the roots. Therefore, experiments were performed, in which the internal production of glutathione was enhanced by fumigation with sulphur dioxide or hydrogen sulphide (Herschbach, 1992). Fumigation of the shoots of tobacco and spinach plants with these sulphur gases reduced sulphate influx and the percentage of the sulphate taken up that was loaded into the xylem of the roots. However, the effects were highly dependent on sulphur nutrition of the plants. From these experiments with cultured cells, excised roots and intact plants it appears that a sulphur compound produced in the leaves and transported to the roots can – to some extent – prevent the uptake of excess sulphate by the roots. Glutathione seems to be the likely candidate for this compound (Fig. 2).

Membrane transport of glutathione

Distribution of glutathione produced in mature leaves within the whole plant by phloem and xylem transport can only be achieved, when glutathione is loaded into, and unloaded from the long-distance transport tissues by membrane transport. Export of glutathione has been observed in bacterial (Owens & Hartman, 1986), animal (Meister & Anderson, 1983), and plant cells (Rennenberg, 1982). In green tobacco suspension cultures, e.g. up to 99% of the glutathione produced by the cells is exported in the form of GSH into the culture medium (Bergmann & Rennenberg, 1978; Rennenberg & Bergmann, 1979). Little, if any, intact glutathione is transported into animal tissue (Meister & Anderson, 1983). External glutathione is degraded by membrane-bound γ-glutamyl transpeptidase and the reaction products are taken up by the cells. Experiments with growing tobacco cells indicate that intact glutathione can be transported into plant tissue (Rennenberg, 1981). When the cells were exposed to radiolabelled glutathione in the presence of inhibitors of glutathione synthesis, radioactivity inside the cells could primarily be attributed to glutathione. Apparently, transport of external glutathione is mediated by different processes in animal and plant tissue.

This assumption was confirmed by recent membrane transport experiments with cultured tobacco cells (Schneider, Martini & Rennenberg, 1992). Uptake of GSH into both green and heterotrophic tobacco cells proceeds with biphasic Michaelis–Menten kinetics in the concentration range of 0.01–1.0 mM. For the high affinity system K_m-values were similar for both type of cells (17–18 µM). In heterotrophic cells the low affinity system showed a higher affinity for GSH (K_m 310 µM) than in green cells (K_m 780 µM). Maximum transport velocities for the low (167–177 nmol GSH min^{-1} g^{-1} dry weight) and the high affinity system (19–20 nmol GSH min^{-1} g^{-1} dry weight) were in the same range for heterotrophic and green tobacco cells. Maximum rates of uptake were obtained at pH 5.5–6.5. GSH uptake was found to be an active, energy driven process. γ-Glutamyl transpeptidase is unlikely to be involved in GSH uptake by tobacco cells (Schneider et al., 1992). L-Cysteine was identified as a competitive inhibitor of GSH uptake by the high affinity transport system of both heterotrophic and green tobacco cells. Other amino acids did not show a considerable effect on GSH transport. The presence of sulphate diminished GSH transport into green, but not into heterotrophic cells (Schneider et al., 1992). This finding confirms observations with growing suspension cultures, where sulphate was found to inhibit GSH uptake by green, but not by heterotrophic cells (Rennenberg et al., 1988). Factors controlling membrane transport of GSH in loading and unloading of phloem and xylem have not been identified so far.

Conclusions

The glutathione concentration in plant cells is determined by a whole set of processes (Fig. 3). In mature leaves, synthesis and degradation of glutathione, its use in biosynthetic processes and export of glutathione to other plant organs (including loading of glutathione into transport tissues and its long-distance transport) modulate the concentration of glutathione. In heterotrophic tissues and developing leaves, synthesis of glutathione may be of minor impact for the concentration of glutathione. In these tissues, glutathione degradation, its use in biosynthetic processes, and import of glutathione from phloem and xylem (including long-distance transport and unloading of glutathione) determine the glutathione content. At present, the significance of individual processes for the concentration of glutathione in plant cells

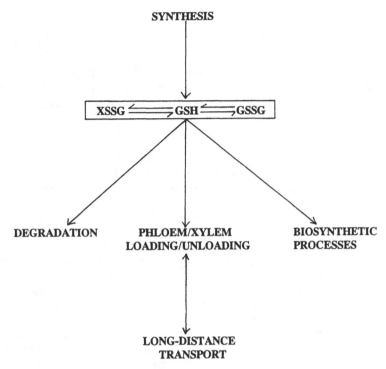

Fig. 3. Processes modulating the concentration of glutathione in plant cells (modified from: Rennenberg & Lamoureux, 1990). GSH, reduced glutathione; GSSG, glutathione disulphide; XSSG, glutathione mixed disulphide.

cannot be properly estimated, because the fluxes of sulphur through these processes have not been determined. An exception are first labelling studies, in which the rate of long-distance transport of gluta- thione has been calculated (Rauser *et al.*, 1991; Schupp *et al.* 1992). Feedback inhibition of glutathione on its own synthesis, and gluta- thione-mediated inhibition of sulphate influx and xylem loading of sulphate in the roots, are the only regulatory process of glutathione metabolism that have been studied in plant cells to some extent. Obviously, many other regulatory processes are required to obtain an optimum distribution of sulphur within the plant during growth and development by glutathione synthesis and transport. Information on these processes seems also to be essential for a proper interpretation of results obtained in studies on the role of glutathione in stress physiology.

Acknowledgements

Work by the author's group reported here was financially supported by the Deutsche Forschungsgemeinschaft (DFG) and the Bundesminister für Forschung und Technologie (BMFT). Fumigation experiments with sulphur dioxide or hydrogen sulphide were performed in cooperation with Dr Luit J. De Kok, University of Groningen.

References

Alosi, M.C., Melroy, D.L. & Park, R.B. (1988). The regulation of gelation of phloem exudate from *Cucurbita* fruit by dilution, glutathione, and glutathione reductase. *Plant Physiology*, **86**, 1089–94.

Bergmann, L. & Rennenberg, H. (1978). Efflux und Produktion von Glutathion in Suspensionskulturen von *Nicotiana tabacum*. *Zeitschrift für Pflanzenphysiologie*, **88**, 175–85.

Bergmann, L. & Rennenberg, H. (1993). Glutathione metabolism in plants. In *Sulfur Nutrition and Sulfur Assimilation in Higher Plants*, ed. L.J. De Kok, I. Stulen, H. Rennenberg, C. Brunold & W. Rauser, pp. 109–23. The Hague: SPB Acad. Publ.

Bonas, U., Schmitz, K., Rennenberg, H. & Bergmann, L. (1982). Phloem transport of sulphur in *Ricinus*. *Planta*, **155**, 82–8.

Brunold, C. (1990). Reduction of sulfate to sulfide. In *Sulfur Nutrition and Sulfur Assimilation in Higher Plants*, ed. H. Rennenberg, C. Brunold, L.J. De Kok & I. Stulen, pp. 13–31. The Hague: SPB Acad. Publ.

Buwalda, F., De Kok, L.J., Stulen, I. & Kuiper, P.J.C. (1988). Cysteine, γ-glutamylcysteine and glutathione contents of spinach leaves as affected by darkness and application of excess sulphur. *Physiologia Plantarum*, **74**, 663–8.

Buwalda, F., Stulen, I., De Kok, L.J. & Kuiper, P.J.C. (1990). Cysteine, γ-glutamylcysteine and glutathione contents of spinach leaves as affected by darkness and application of excess sulphur. II. Glutathione accumulation in detached leaves exposed to H_2S in the absence of light stimulated by the supply of glycine to the petioles. *Physiologia Plantarum*, **80**, 196–204.

De Kok, L.J. (1990). Sulfur metabolism in plants exposed to atmospheric sulfur. In *Sulfur Nutrition and Sulfur Assimilation in Higher Plants*, ed. H. Rennenberg, C. Brunold, L.J. De Kok & I. Stulen, pp. 111–30. The Hague: SPB Acad. Publ.

Garsed, S.G. & Read, D.J. (1977a). Sulphur dioxide metabolism in soy-bean, *Glycine max* var. biloxi. I. The effects of light and dark on uptake and translocation of $^{35}SO_2$. *New Phytologist*, **78**, 111–19.

Garsed, S.G. & Read, D.J. (1977b). Sulphur dioxide metabolism in soy-bean, *Glycine max* var. biloxi. II. Biochemical distribution of $^{35}SO_2$ products. *New Phytologist*, **79**, 583–92.

Giovanelli, J. (1990). Regulatory aspects of cysteine and methionine synthesis. In *Sulfur Nutrition and Sulfur Assimilation in Higher Plants*, ed. H. Rennenberg, C. Brunold, L.J. De Kok & I. Stulen, pp. 33–48. The Hague: SPB Acad. Publ.

Gojon, A., Soussana, J.-F., Passama, L. & Robin, P. (1986). Nitrate reduction in roots and shoots of barley (*Hordeum vulgare* L.) and corn (*Zea mays* L.) seedlings. I. ^{15}N Study. *Plant Physiology*, **82**, 254–60.

Gunz, G., Herschbach, C. & Rennenberg, H. (1992). Specificity of glutathione (GSH) mediated inhibition of sulphate uptake into excised tobacco roots. *Phyton*, **32**, 51–4.

Hatzios, K.K. & Hoagland, R.E. (1989). *Herbicide Safeners: Progress and Prospects*. New York: Academic Press

Herschbach, C. (1992). Untersuchungen zur Bedeutung von Glutathion (GSH) für die 'inter-organ' Regulation der Sulfatversorgung an Tabakpflanzen (*Nicotiana tabacum* L. var. 'Samsun'). PhD Thesis, Technical University Munich.

Herschbach, C. & Rennenberg, H. (1991). Influence of glutathione (GSH) on sulphate influx, xylem loading and exudation in excised tobacco roots. *Journal of Experimental Botany*, **42**, 1021–9.

Lamoureux, G.L. & Rusness, D.G. (1989). The role of glutathione and glutathione-S-transferases in pesticide metabolism, selectivity, and mode of action in plants and insects. In *Glutathione: Chemical, Biochemical and Medical Aspects*, vol. IIIB, ed. D. Dolphin, R. Poulson & O. Awamovic, pp. 153–96. New York: Wiley & Sons Publ.

Macnicol, P.K. & Bergmann, L. (1984). A role of homo-glutathione in organic sulphur transport to the developing mung bean seed. *Plant Science Letters*, **53**, 229–35.

Meister, A. (1981). On the cycles of glutathione metabolism and transport. In *Current Topics in Cellular Regulation*, vol. 18, *Biological Cycles*, ed. R.W. Estabrook & P. Srere, pp. 21–58. New York: Academic Press.

Meister, A. & Anderson, M.E. (1983). Glutathione. *Annual Review of Biochemistry*, **52**, 711–60.

Owens, R.A. & Hartman, P.E. (1986). Export of glutathione by some widely used *Salmonella typhimurium* and *Escherichia coli* strains. *Journal of Bacteriology*, **168**, 109–14.

Polle, A. & Rennenberg, H. (1993). Significance of antioxidants in plant adaptation to environmental stress. In *Plant Adaptation to Environmental Stress*, ed. L. Fowden, T. Mansfield & F. Stoddard, pp. 263–73. London: James & James.

Polle, A. & Rennenberg, H. (1994). Photooxidative stress in trees. In *Causes of Photooxidative Stress and Amelioration of Defense Systems in Plants*, ed. C. Foyer & P. Mullineaux, pp. 199–218. Boca Raton: CRC Press

Rauser, W.E., Schupp, R. & Rennenberg, H. (1991). Cysteine, γ-glutamylcysteine and glutathione levels in maize seedlings: distribution and translocation in normal and Cd exposed plants. *Plant Physiology*, **97**, 128–38.

Rennenberg, H. (1981). Differences in the use of cysteine and glutathione as sulphur source in photoheterotrophic tobacco suspension cultures. *Zeitschrift für Pflanzenphysiologie*, **105**, 31–40.

Rennenberg, H. (1982). Glutathione metabolism and possible biological roles in higher plants. *Phytochemistry*, **21**, 2771–81.

Rennenberg, H. (1984). The fate of excess sulphur in higher plants. *Annual Review of Plant Physiology*, **35**, 121–53.

Rennenberg, H. & Bergmann, L. (1979). Einfluss von Ammonium und Sulfat auf die Glutathion-Produktion in Suspensionskulturen von *Nicotiana tabacum*. *Zeitschrift für Pflanzenphysiologie*, **92**, 133–42.

Rennenberg, H. & Brunold, C. (1994). Significance of glutathione metabolism in plants under stress. *Progress in Botany*, **55**, 142–56.

Rennenberg, H. & Lamoureux, G.L. (1990). Physiological processes that modulate the concentration of glutathione in plant cells. In *Sulfur Nutrition and Sulfur Assimilation in Higher Plants*, ed. H. Rennenberg, C. Brunold, L.J. De Kok & I. Stulen, pp. 53–65. The Hague: SPB Acad. Publ.

Rennenberg, H., Kemper, O. & Thoene, B. (1989). Recovery of sulphate transport into heterotrophic tobacco cells from inhibition by reduced glutathione. *Physiologia Plantarum*, **76**, 271–6.

Rennenberg, H. & Polle, A. (1994). Metabolic consequences of atmospheric sulphur influx into plants. In *Plant Responses to the Gaseous Environment*, ed. R.G. Alscher & A.R. Wellburn, pp. 165–80, London: Chapman & Hall.

Rennenberg, H., Polle, A., Martini, N. & Thoene, B. (1988). Interaction of sulphate and glutathione transport in cultured tobacco cells. *Planta*, **176**, 68–74.

Rennenberg, H., Schmitz, K. & Bergmann, L. (1979). Long-distance transport of sulphur in *Nicotiana tabacum*. *Planta*, **147**, 57–62.

Rennenberg, H. & Thoene, B. (1987). The mobility of sulphur in higher plants. In *Proceedings of the International Symposium Elemental Sulphur in Agriculture*, vol. 2, pp. 701–7. Nice.

Rüegsegger, A. & Brunold, C. (1992). Effect of cadmium on γ-glutamylcysteine synthesis in maize seedlings. *Plant Physiology*, **99**, 428–33.

Schneider, A. (1992). Untersuchungen zum Glutathion-Transport und -Stoffwechsel in Suspensionskulturen von *Nicotiana tabacum* L. var. 'Samsun'. PhD Thesis, Technical University Munich.

Schneider, A., Martini, N. & Rennenberg, H. (1992). Reduced glutathione (GSH) transport into cultured tobacco cells. *Plant Physiology and Biochemistry*, **30**, 29–38.

Schneider, A., Kreuzwieser, J., Schupp, R., Sauter, J.J. & Rennenberg, H. (1994a). Thiol and amino acid composition of the xylem sap of poplar trees (Populus × canadensis 'robusta'). *Canadian Journal of Botany*, **72**, 347–51.

Schneider, A., Schatten, T. & Rennenberg, H. (1994*b*). Exchange between phloem and xylem during long distance transport of glutathione in spruce trees (*Picea abies* [Karst.] L.). *Journal of Experimental Botany*, **45**, 457–62.

Schupp, R. (1991). Untersuchungen zur Schwefelernährung der Fichte (*Picea abies* L.): Die Bedeutung der Sulfatassimilation und des Transportes von Thiolen. PhD Thesis, Technical University Munich.

Schupp, R. & Rennenberg, H. (1992). Changes in sulphur metabolism during needle development of Norway spruce. *Botanica Acta*, **105**, 180–9.

Schupp, R., Glavac, V. & Rennenberg, H. (1991). Thiol composition of xylem sap of beech trees. *Phytochemistry*, **30**, 1415–18.

Schupp, R., Schatten, T., Willenbrink, J. & Rennenberg, H. (1992). Long-distance transport of reduced sulphur in spruce (*Picea abies* L.). *Journal of Experimental Botany*, **43**, 1243–50.

Schütz, B., De Kok, L.J. & Rennenberg, H. (1991). Thiol accumulation and cysteine desulfhydrase activity in H_2S-fumigated leaves and leaf homogenates of cucurbit plants. *Plant and Cell Physiology*, **32**, 733–6.

Smith, I.K., Polle, A. & Rennenberg, H. (1990). Glutathione. In *Stress Responses in Plants: Adaptation and Acclimation Mechanisms*, ed. R. Alscher & I. Cunning, pp 201–15. New York: Wiley-Liss.

Touraine, B., Grignon, N. & Grignon, C. (1990). Interaction between nitrate assimilation in shoots and nitrate uptake by roots of soybean (*Glycine max*) plants: role of carboxylate. In *Plant Nutrition – Physiology and Applications*, ed. M.L. van Beusichem, pp. 87–92. Dordrecht: Kluwer Academic Publishers.

JOHN GORHAM

Betaines in higher plants – biosynthesis and role in stress metabolism

Introduction

Betaines are quaternary ammonium compounds (QACs) which contain a carboxylic acid group. They may generally be regarded as fully N-methylated amino or imino acids. While the most commonly encountered betaine (glycinebetaine) is structurally, although not biosynthetically, based on the simplest amino acid, a wide range of aliphatic and aromatic betaines have been described (Fig. 1). These include imino and aromatic amino acid betaines, and compounds such as trigonelline (nicotinic acid betaine) and homostachydrine (pipecolic acid betaine). Related compounds such as choline-O-sulphate and the tertiary sulphonium analogue of β-alaninebetaine, 3-dimethylsulphoniopropionate (DMSP), are also considered in this review since they may have similar physiological roles in some plants.

Although the distribution of betaines in plants is imperfectly known, partly because many taxa have not been thoroughly investigated, it is clear that some families normally contain significant amounts (> 10 mol m^{-3} plant water) of betaines. In lower plants betaines and related compounds may be incorporated into lipids (Sato, 1992). Recent developments in mass and NMR spectroscopy have greatly facilitated the qualitative and quantitative analysis of betaines. Molecular biology has provided useful tools for the study of regulation of key enzymes of betaine biosynthesis. *In vitro* studies have clarified the role of betaines as compatible solutes in plants subjected to a variety of stresses. Developments in these areas since the review by Wyn Jones and Storey (1981) will be discussed in detail here, and have also been reviewed by Rhodes and Hanson (1993).

Distribution of betaines and related compounds in plants

Many plants contain trace amounts of glycinebetaine, but the accumulation of larger concentrations of this or other betaines is a feature of

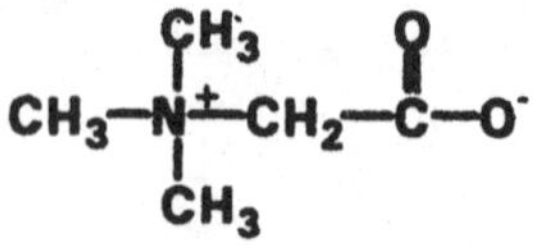

GYCINEBETAINE
(LYCINE)

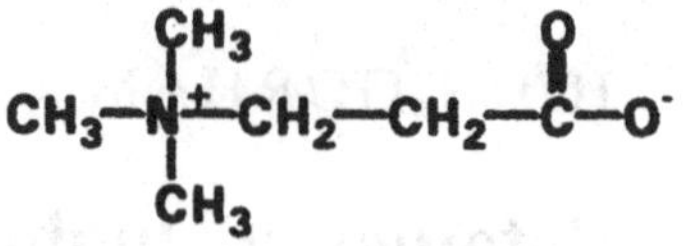

β-ALANINEBETAINE
(HOMOBETAINE)

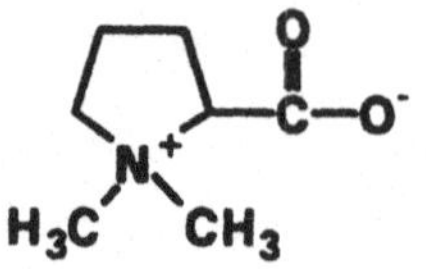

STACHYDRINE
(PROLINEBETAINE)

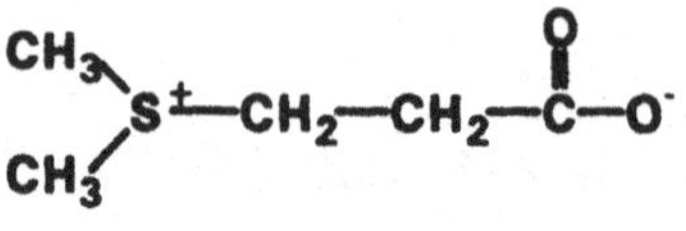

3-DIMETHYLSULPHONIOPROPIONATE

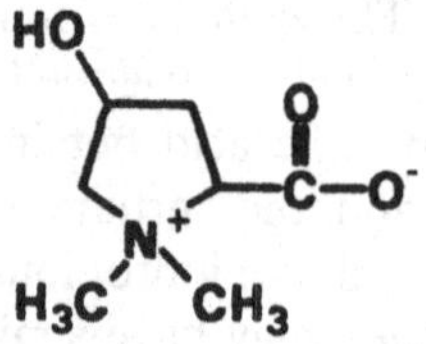

4-HYDROXYPROLINEBETAINE
(BETONICINE)

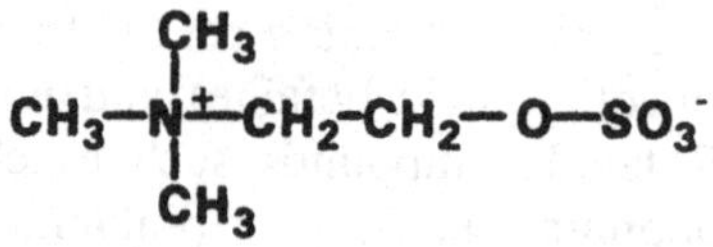

CHOLINE-O-SULPHATE

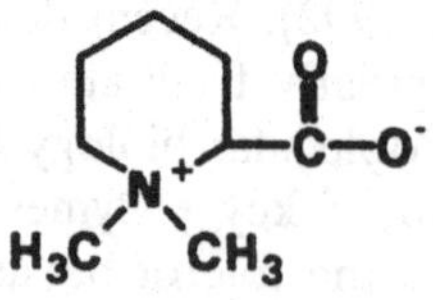

PIPECOLATEBETAINE
(HOMOSTACHYDRINE)

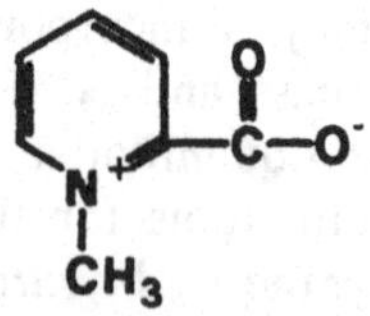

NICOTINIC ACID BETAINE
(TRIGONELLINE)

Fig. 1. Structures of plant betaines.

relatively few taxa. The known distribution of betaine accumulation in angiosperm orders (as defined by Takhtajan, 1969) is illustrated in Fig. 2. The application of mass and NMR spectrometry to further taxa should help to clarify what at the moment is a very incomplete picture. Lack of betaine accumulation, as indicated in Fig. 1, is as much a reflection of the absence of systematic studies as of the absence of the biosynthetic pathways. Weretilnyk *et al.* (1989) suggested that the glycinebetaine pathway may be widely distributed in higher plants, but in many cases only weakly expressed. This was supported by the finding of low concentrations (about 0.4 μmol g^{-1} DW) of glycinebetaine in the primitive angiosperm *Magnolia x soulangiana* by fast atom bombardment mass spectrometry (FABMS) and desorption chemical ionization mass spectrometry (DCIMS). Activity of the glycinebetaine-synthesizing enzyme betaine aldehyde dehydrogenase (BADH) was below the level of detection in assays or on native polyacrylamide gels, but a protein which cross-reacted with antibodies to *Spinacea oleracea* BADH was found by immunoblotting.

In the monocots, glycinebetaine is widely, although not universally, distributed in the Poales. The principal dicot families in which glycinebetaine accumulation is common are the Asteraceae (Asterales), Chenopodiaceae, Caryophyllaceae and Amaranthaceae (Caryophyllales), Solanaceae (Scrophulariales), Malvaceae (Malvales), Convolvulaceae and Cuscutaceae (Polemoniales), Avicenniaceae (Lamiales) Plumbaginaceae (Plumbaginales), Fabaceae (Fabales) and Asteraceae (Asterales) (Wyn Jones & Storey, 1981; Rhodes & Hanson, 1993; Weretilnyk *et al.*, 1989). Recent investigations of quaternary ammonium compounds in the Malvaceae revealed the widespread distribution of glycinebetaine in this family, although two species in which glycinebetaine was not detected contained another, unidentified QAC (Table 1).

β-Alaninebetaine and choline-O-sulphate are found particularly in the Plumbaginaceae (Hanson *et al.*, 1991). In this family choline-O-sulphate was found in all species examined, but the distribution of glycinebetaine and β-alaninebetaine was mutually exclusive. The former was found in four genera of the subfamily Plumbagoideae, in *Gladiolimon* and in species of *Limonium* belonging to the section Pteroclados (Hanson, personal communication). With the exception of two prolinebetaine-containing species in the section Myriolepis of the genus *Limonium*, all other species accumulated β-alaninebetaine. Three species, one of *Arthrolimon* and two of *Limoniastrum*, contained both prolinebetaine (stachydrine) and an hydroxyprolinebetaine. Choline-O-sulphate may have a role in sulphur transport, and it may be associated with the presence of salt glands.

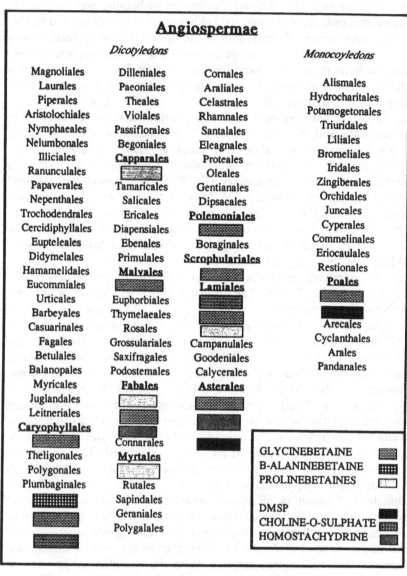

Fig. 2. Distribution of betaines in higher plants.

Table 1. *Distribution of quaternary ammonium compounds in the Malvaceae. Compounds were detected with Dragendorff's reagent on silica gel thin-layer plates developed in methanol:acetone:HCl (90:10:4).*

Taxon	choline	glycinebetaine	?
Abelmoschatus moschatus	++	−	++
Abutilon hybridum maximum	++	++	++
A.vitifolium cv. Veronica Tenant	+	++	−
Althaea ficifolia	++	++	−
A.officinalis	++	++	−
A.rosea cv. Majorette	++	++	++
A.rosea var. *nigra*	++	++	+
Anoda cristata cv.Opal Cup	++	++	−
A.cristata cv.Silver Cup	++	++	−
Gossypium anomalum	+	++	−
G.armourianum	+	++	-
G.australe	+	++	−
G.bickii	-	++	−
G.capitis-viridis	+	++	-
G.costulatum	-	++	−
G.davidsonii	−	++	−
G.harknessii	+	++	−
G.klotzschianum	−	++	-
G.longicalyx	+	++	−
G.nelsonii	++	++	−
G.raimondii	−	++	?
G.sturtianum	+	++	−
G.thurberi	++	++	−
G.tomentosum	−	++	−
G.trilobum	+	++	−
Hibiscus coccineus	++	++	+
H.diversifolius	++	++	−
H.moschatus	++	++	+
H.mutabilis	++	++	+
H.syriacus	+	++	−
Hibiscus cv.Southern Belle	+	++	−
Hibiscus sp.	+	++	−
Hibiscus sp.	+	−	+
Kitaibelia vitifolia	++	++	−
Lavatera cretica	++	++	++
L.trimestris cv.Silver Cup	++	++	−
Malope trifida cv.Vulcan	+	++	−
Malva alcea var.*fastigiata*	++	++	−
M.crispa	++	++	−
M.moschata	++	++	−
M.moschata var.*alba*	++	++	−
M.neglecta	++	++	−
M.sylvestris	++	++	−

The Fabaceae is another family in which a variety of betaines have been found. These include glycinebetaine, prolinebetaine, hydroxyprolinebetaine and pipecolatebetaine (homostachydrine), but they do not occur in all species. The genus *Medicago*, including the forage crop alfalfa (*M. sativa*), has been studied in some detail (Naidu, Paleg & Jones, 1992; Wood *et al.*, 1991). Prolinebetaine was the major betaine in alfalfa, but the relative amounts of the other betaines were genotype-dependent (Wood *et al.*, 1991). Analyses of red clover (*Trifolium pratense*) by the same methods (FABMS and DCIMS) showed that glycinebetaine was the main betaine in this species. Prolinebetaine was predominant in birdsfoot trefoil (*Lotus corniculatus*). Of seven annual species of *Medicago*, all contained glycinebetaine, none contained pipecolatebetaine and five contained prolinebetaine (Naidu *et al.*, 1992). *M. sativa* cv. Hunter River contained prolinebetaine and pipecolatebetaine, but no glycinebetaine. Small amounts of trigonelline were found in most species.

Prolinebetaine and hydroxyprolinebetaines (3-hydroxy- and 4-hydroxy-) have also been found in *Batis, Achillea, Lamium* and *Stachys,* and pipecolatebetaine was found in *Achillea* (Rhodes & Hanson, 1993). 4-hydroxypipecolate betaine occurs in *Lamium maculatum* (Yuan *et al.*, 1992). Prolinebetaine was reported to occur in several members of the Labiateae (Wyn Jones & Storey, 1981), and hydroxyprolinebetaines were found in *Melaleuca* species (Jones *et al.*, 1987; Naidu, *et al.*, 1987). Both betaines also occur in the family Capparaceae (Delaveau Kondogbo & Pousset, 1973; Cornforth & Henry, 1952). A summary of the distribution of prolinebetaine in plants is presented in Table 2.

Nicotinic acid betaine (trigonelline) is found in small quantities in most plants, but rarely accumulates to osmotically significant concentrations. The tertiary sulphonium compound, DMSP, which is well known from a variety of marine algae (Blunden & Gordon, 1986; Blunden *et al.*, 1992) has been found in a few higher plants. It was first reported from *Spartina anglica* (Poaceae) by Larher, Hamelin and Stewart (1977), and more recently from *Melanthera (Wedelia) biflora* (Asteraceae) by Storey *et al.* (1993) and *Saccharum* (Poaceae) by Paquet *et al.* (1994). Different proportions of DMSP and glycinebetaine were found in different accessions of *Melanthera biflora* (Storey *et al.*, 1993). Hanson, Huang and Gage (1993) have presented convincing evidence that for the 5-dimethylsulphoniopentanoate reported (Larher and Hamelin, 1979) to occur in *Diplotaxis tenuifolia* is an artefact of extraction.

Table 2. *Occurrence of prolinebetaine (stachydrine) in higher plants.*

Poaceae	**Rutaceae**
Oryza sativa	*Citrus* spp.
Capparidaceae	**Moraceae**
Capparis tomentosa	*Cudrania javanensis*
C. spp.	
Cadaba fruticosa	**Fabaceae**
Labiateae	*Desmodium* spp.
	Erythrina variegata
Marrubium vulgare	Medicago spp.
Lagochilus spp.	*Lotus corniculatus*
Galeopsis ochroleuca	
Leonurus spp.	**Plumbaginaceae**
Stachys spp.	*Myriolepis* spp.
Lamium album	*Arthrolimon* sp.
Eremostachys speciosa	*Limoniastrum* spp.
Sideritis montana	
Asteraceae	**Bataceae**
Onopordium acanthium	*Batis* sp.
O. alexandrinum	
Chrysanthemum cinerariaefolium	
C. sinense	
Achillea spp.	

For references see Delaveau *et al.* (1973), McLean (1976), Rhodes & Hanson (1993) and Wyn Jones & Storey (1981).

Examples of other betaines are isoleucinebetaine from *Cannabis sativa* (Bercht *et al.*, 1973), DOPA-betaine from *Lobaria laetivirens* (Bernard *et al.*, 1981), tyrosinebetaine (maokonine) from *Lobaria laetivirens* (Bernard *et al.*, 1981), *Ephedra* sp. (Tamada, Endo & Hikino, 1978) and several species in the Moraceae (Okogun, Spiff & Ekong, 1976; Shrestha & Bisset, 1991) and phenylalaninebetaine and tryptophanbetaine (hypaphorine) from *Antiaris africana* (Okogun, Spiff & Ekong, 1976).

Analytical methods

Progress in the analytical techniques, both quantitative and qualitative, for betaines has lagged behind those for most biological compounds for two reasons. The ionic nature of the molecule and the lack of reactivity of the trimethylamino group render betaines insufficiently volatile for conventional gas-liquid chromatography (GLC), although pyrolysis-GLC has been used successfully for the analysis of glycinebetaine (Hitz & Hanson, 1980). The second problem, at least for aliphatic betaines, is the lack of a chromophore other than the carboxylic acid group. Detection by UV absorbance in HPLC is thus limited to low (< 210 nm) wavelengths, leaving refractive index detection as the most practical, although less sensitive, alternative.

Of the various HPLC methods which have been employed for the separation of betaines, the most generally useful employ carbohydrate columns (indeed some of the methods were developed for the beet sugar industry, and to detect betaine, as an impurity in beet sugar, in wine). These carbohydrate columns may be either amino-bonded silica columns eluted with water–acetonitrile mixtures (Steinle & Fischer, 1978; Vialle, Kolosky & Rocca, 1981) or cation exchange columns (usually in the Ca^{2+} or Na^+ forms) eluted with water (Guy, Warne & Reid, 1984; Rajakulä & Paloposki, 1983; Wolff et al., 1989). Both carbohydrates and betaines may be determined on these columns, and retention of betaines on the cation exchange columns can be adjusted by the addition of the appropriate cation (Ca^{2+} or Na^+) to the eluant. For studies with salt-treated plant material, Na^+-form columns are easier to maintain in the desired ionic form. Betaines may also be separated on strong cation exchange (sulphonic acid-bonded) silica (Gorham, 1984). The problem of detection may be overcome by the use of UV-absorbing derivatives such as p-bromophenacyl esters of betaines, which can be separated on weak cation exchange (carboxylic acid-bonded) silica columns (Gorham, 1986; Minkler et al., 1984). Glycinebetaine aldehyde and choline have also been separated as the p-nitrobenzyl oxime and the benzoyl or dinitrobenzoyl esters respectively (Gorham & McDonnell, 1985).

Both ^1H and ^{13}C NMR spectroscopy are useful tools for the quantitative and qualitative analysis of betaines (Blunden et al., 1986; Chastellain & Hirsbrunner, 1976; Jones et al., 1986; Larher, 1988). With the addition of suitable standards, FABMS and DCIMS provide the most sensitive, and structurally informative, analytical methods. FABMS involves the analysis of an ester (n-butyl or n-propyl), or the cation adduct of the free betaine or related compound, in glycerol (Hanson & Gage, 1991; Rhodes et al., 1987). DCIMS uses the underivatized betaines coated on a rhenium filament (Wood et al., 1991).

The chromatographic and spectrometric methods described above require some prior purification of plant extracts. Fortunately, betaines have characteristic interactions with ion exchange materials which allow rapid separation of betaines from other plant components (Rhodes & Hanson, 1993). Separation of a betaine fraction is also desirable prior to simpler analyses on thin-layer plates and detection with Dragendorff's reagent (Blunden *et al.*, 1981; Gorham *et al.*, 1981). A routine method for the estimation of QACs in crude extracts involves precipitation of periodides (Barak & Tuma, 1979; Grieve & Grattan, 1983; Storey & Wyn Jones, 1977), although betaines may be overestimated by this technique. Preliminary purification is also necessary before determination of betaines by UV spectrometry of *p*-bromophenacyl esters (Gorham, McDonnell & Wyn Jones, 1982).

Accumulation of betaines in stressed plants

One of the main reasons for the interest in betaines generally, and glycinebetaine in particular, is their suggested role as 'compatible solutes', a term which will be discussed in more detail later. Consistent with this role, glycinebetaine (and some other betaines) accumulate to higher concentrations in some plants in response to a variety of environmental stresses. This response is normally seen in those plants which already contain some (> 1–2 mol m^{-3} plant water) glycinebetaine in the unstressed condition. In some cases increased activity of the biosynthetic pathway is clearly responsible for glycinebetaine accumulation (e.g. in spinach, Coughlan & Wyn Jones, 1982; Pan, 1983; Weigel, Weretilnyk & Hanson, 1986). In other cases, where the accumulation is less, it will be necessary to confirm, by enzyme assays or molecular probes, that specific pathways are activated. An alternative possibility is that glycinebetaine accumulation is the result of reduced growth (i.e. less dilution) and increased availability of substrate, especially where growth is inhibited more than photosynthesis.

The accumulation of betaines in response to salt stress has been studied extensively in the Chenopodiaceae and the Poaceae, and much of the earlier work was reviewed by Wyn Jones & Storey (1981). Guy *et al.* (1984) illustrated the problem introduced by using different bases for expressing betaine concentrations. In the succulent chenopod *Salicornia europaea* (which shows the typical chenopod response of increased growth at moderate salinities) glycinebetaine concentrations decreased on a dry weight basis with increasing salinity, but increased significantly when expressed per gram organic matter. Since the main interest here is the osmotic contribution of glycinebetaine, these units are not entirely appropriate. Recalculation of the data of Guy *et al.* (1984) confirms the increased osmotic contribution when the concentrations are expressed on a plant water basis. Salt-induced glycinebeta-ine accumulation has been demonstrated in cell suspension cultures as well as intact plants of *Atriplex semibaccata* and *A. halimus* (Koheil *et al.*, 1992). Higher glycinebetaine concentrations were found in suspension cultures treated with elevated concentrations of macronutrients, indicating a similar response to osmotic and salt stresses. Glycinebetaine accumulation is feature of both salt-tolerant and glycophytic species of the Chenopodiaceae (Weigel & Larher, 1985). In the related family Ameranthaceae, water stress induces glycinebetaine accumulation in *Ameranthus hypochondriacus* (Gamboa, Valenzuela & Murillo, 1991).

Glycinebetaine accumulation in salt-stressed cereals is well documented, although different species differ in the extent of accumulation. Basal and adapted concentrations are somewhat higher in barley than in wheat (Murray & Ayres, 1986; Hitz & Hanson, 1980; Araya *et al.*, 1991; Grieve & Maas, 1984; Grattan & Grieve, 1985), and higher in sorghum than in maize (Brunk, Rich & Rhodes, 1989; Grieve & Grattan, 1983; Grieve & Maas, 1984; Rhodes *et al.*, 1987,1989,1993). Cold stress also induces the accumulation of glycinebetaine and other nitrogenous solutes in wheat seedlings (Naidu *et al.*, 1991).

Hanson and co-workers (Hanson & Nelson, 1978; Hanson & Scott, 1980; Hitz & Hanson, 1980; Hitz, Ladyman & Hanson, 1982; Hitz, Rhodes & Hanson, 1981; Ladyman, Hitz & Hanson, 1980; Ladyman *et al.*, 1983) and others (Ahmad & Wyn Jones, 1979; Itai & Paleg, 1982; Zuniga, Argandona & Corcuera, 1989) have thoroughly investigated the accumulation of glycinebetaine in barley in response to water stress. Genetic analysis in barley shows that variation for glycinebetaine accumulation exists between cultivars (and species), and is tightly linked with leaf osmotic potentials (Grumet, Albrechtsen & Hanson, 1987; Grumet & Hanson, 1983,1986; Grumet, Isleib & Hanson, 1985; Ladyman *et al.*, 1983). This linkage prevented an unequivocal answer to the question of the adaptive value of glycinebetaine accumulation in barley, although the results were consistent with such a role. More promising material for experiments to answer this question can be found in maize and sorghum lines deficient in glycinebetaine synthesis (Brunk, Rich & Rhodes, 1989; Lerma *et al.*, 1991; Monyo, Ejeta & Rhodes, 1992; Rhodes *et al.*, 1987, 1989; Rhodes & Rich, 1988; Rhodes & Hanson, 1993). The maize lines differed at a single recessive allele which apparently controlled choline monooxygenase activity.

Drought stress results in glycinebetaine accumulation in some *Medicago* species (Naidu, Paleg & Jones, 1992) and in *Panicum maximum* and *Cenchrus ciliaris* (Ford & Wilson, 1981). In *Medicago* species small increases in prolinebetaine concentrations were also observed.

Glycinebetaine accumulation in citrus seedlings treated with NaCl was observed by Duke, Johnson & Koch (1986). The extent of glycinebetaine accumulation was reduced with increasing mycorrhizal infection. Glycinebetaine accumulation in the Asteraceae has been studied in some detail in *Aster tripolium* subjected to salinity (Gorham, Hughes & Wyn Jones, 1980; Goas, Goas & Larher, 1982; Larher *et al.*, 1982). Accumulation of glycinebetaine has been reported in the Euphorbiaceae (*Euphorbia trigona*) in response to water stress (Huber & Eder, 1982).

In *Limonium* species choline-O-sulphate and glycinebetaine (*L. perezii*) or β-alaninebetaine (*L. latifolium*) concentrations increased with increasing salinity (Hanson *et al.*, 1991). In *L. sinuatum* chloride salinity favoured glycinebetaine synthesis while in sulphate salinity choline-O-sulphate was accumulated to a greater extent. DMSP accumulation in *Spartina* is enhanced by high sulphide, but glycinebetaine concentrations respond more to higher salt and nitrogen levels (Dacey, King & Wakeham, 1987; Cavalieri, 1983; Cavalieri & Huang, 1981; van Diggelen *et al.*, 1986). DMSP has been shown to accumulate in marine algae in response to changes in salinity (e.g. Dickson & Kirst, 1986).

Betaines as compatible and protective solutes

Glycinebetaine was proposed as a compatible solute (i.e. an organic solute which does not inhibit enzyme activity or other metabolic processes and is preferentially accumulated in the cytoplasm in stressed plants) by Wyn Jones *et al.* in 1977. Evidence for compatibility with a number of enzymes was presented by Pollard & Wyn Jones (1979). Preferential location in the cytoplasm was demonstrated by Hall, Harvey & Flowers (1978) using iodoplatinate staining in *Suaeda maritima*, and by Leigh, Ahmad & Wyn Jones (1981) by analysis of isolated *Beta* vacuoles. Since the vacuole occupies a much larger volume than the cytoplasm, the absolute amounts of glycinebetaine may be higher in the vacuole, but the concentration is higher in the cytoplasm. This was also found to be true for isolated vacuoles of *Atriplex gmelini* (Matoh, Watanabe & Takahashi, 1987). The compatibility of glycinebetaine and related compounds has been established in a number of bacteria (Csonka, 1989; Csonka & Hanson, 1991; Fougere & LeRudulier, 1990*a*,*b*; Gloux & LeRudulier, 1988; Hanson *et al.*, 1991; Imhoff & Rodriguez Valera, 1984; LeRudulier *et al.*, 1983,1984*a*,*b*); Sauvage, Hamelin & Larher, 1983; Strøm *et al.*, 1983).

The reasons for the protective and benign nature of betaines has been studied *in vitro*. The significant properties involve interaction between betaines and macromolecules (Low, 1985; Paleg, Stewart & Bradbeer, 1984; Pollard & Wyn Jones 1979; Somero, 1986; Winzor *et al.*, 1992; Wyn Jones, 1984; Yancey *et al.*, 1982). Inorganic ions tend to destabilize the folding of proteins, whereas betaines and other compatible solutes have the opposite effect. Protection is not only against excessive ion concentrations and osmotic pressures, but also against heat destabilization (Laurie & Stewart, 1990; Nikolopoulus & Manetas, 1991; Nash, Paleg & Wiskich, 1982; Paleg *et al.*, 1981; Shomer Ilan, Jones & Paleg, 1991; Smirnoff & Stewart, 1985). Glycinebetaine prevents loss of activity of phosphoenolpyruvate carboxylase caused by dilution and depolymerization of the enzyme (Manetas, 1990; Nikolo-poulus & Manetas, 1991; Selinioti, Nikolopoulus & Manetas, 1987).

Effects of betaines on membrane stability have also been studied. Rudolph, Crowe & Crowe (1986) examined the interaction of gly-cinebetaine with phospholipids. The effect of glycinebetaine on the heat stability of membranes was reported by Jolivet, Larher & Hamelin (1982), and the same authors (Jolivet, Hamelin & Larher, 1983) found that glycinebetaine and related compounds protected *Beta* membranes from the destabilizing effect of oxalate. At the other end of the temperature scale, betaines have been found to be cryoprotectants for liposomes (Higgins *et al.*, 1987; Lloyd *et al.*, 1992), *Medicago* mem-branes (Zhao, Aspinall & Paleg, 1992), thylakoid membranes (Coughlan & Heber, 1982) and sperm (Koskinen *et al.*, 1989). Ion transport properties of membranes are also affected by glycinebetaine (Ahmad, Wyn Jones & Jeschke, 1987).

Chloroplasts of spinach and *Suaeda* contain high concentrations of glycinebetaine (Génard *et al.*, 1991; Robinson & Jones, 1986; Schröp-pel-Meier & Kaiser, 1988), and this compound has been shown to protect various aspects of photosynthesis, including the functioning of ribulose bisphosphate carboxylase-oxygenase (Incharoensakdi, Takobe & Akazawa, 1986; Murata *et al.*, 1992; Papageorgiou, Fuji-mura & Murata, 1991; Williams, Brain & Dominy, 1992). The effects of glycinebetaine on RNA translation and RNAase activity have been reported by Brady *et al.* (1984), Billard, Rouxel & Boucard (1986) and Gibson, Speirs & Brady (1984).

Determining the compatibility of DMSP and β-alaninebetaine is com-plicated by the instability of these β-trimethylamino substituted acids in the presence of alkaline metal ions at alkaline pH. DMSP was found to be effective in replacing the osmoprotective role of glycinebetaine in *Escherichia coli* (Chambers *et al.*, 1987), and β-alaninebetaine increased the growth of *E. coli* and *Salmonella typhimurium* in the presence of 600 mol m^{-3} NaCl (Hanson *et al.*, 1991). Enzymes from the alga *Tetraselmis subcordiformis* were relatively unaffected by high concentrations of DMSP (Gröne & Kirst, 1991).

It is of some interest to compare the responses of two nitrogen-containing compatible solutes, proline and glycinebetaine, to salt stress. The rate and extent of accumulation are influenced by the rate of application of the stress, be it salt (Gorham, unpublished observations) or water (Naidu *et al.*, 1990) stress. In the case of salt, a gradual increase in stress induces betaine accumulation in wheat and cotton. Proline does not begin to accumulate in these species until the capacity of the plant to adjust to the salt stress is exceeded, in the sense that osmotic adjustment either fails to avoid a simultaneous water stress or until ion concentrations within the plant become toxic.

An interesting interaction between plants and bacteria is seen in nitrogen-fixing root nodules of legumes. Betaines protect the bacteria, as well as the host, from the detrimental effects of salt or osmotic stress (Bouillard & Le Rudulier, 1983; Pocard, Bernard & Le Rudulier, 1991; Pocard *et al.*, 1984; Pocard & Smith, 1988; Smith *et al.*, 1988).

Several attempts have been made to demonstate the beneficial effects of glycinebetaine on stressed plants by exogenous application of the betaine. Many of these studies are difficult to interpret since soil and leaf microorganisms readily take up and metabolise glycinebetaine. Effects on water use efficiency in wheat (Bergmann & Eckert, 1984) and freezing tolerance in alfalfa (Zhao, Aspinall & Paleg, 1992) have been reported. Axenic cultures of barley embryos were protected from the effects of salt stress by 10 mol m^{-3} glcyinebetaine (Lone *et al.*, 1987).

Betaine biosynthesis

Glycinebetaine

Although methylation of glycine might seem the most obvious route to glycinebetaine, biosynthesis in plants does not always take the shortest pathway. In both animals and plants (Cromwell & Rennie, 1954; Delwiche & Bergoff, 1957; Hanson & Nelson, 1978; Hanson & Scott, 1980) ^{14}C-choline is efficiently incorporated into glycinebetaine. The origin of the choline precursor has been studied in some detail (Fig. 3).Tracer experiments (e.g. Coughlan & Wyn Jones, 1982) established serine as the precursor of choline in spinach. Evidence from *Lemna* is consistent with (although not proof of) direct decarboxylation of serine to ethanolamine (Mudd & Datko, 1989). The methylation of ethanolamine may, however, follow a number of different routes involv-

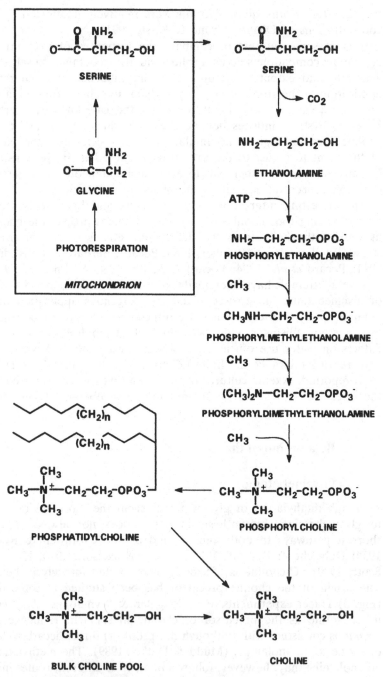

Fig. 3. Pathways from serine to choline.

In glycinebetaine-accumulating plants two pathways have been established. In barley and wheat the main pathway proceeds via phosphoryl intermediates to phosphorylcholine, and thence to phosphatidylcholine (Hitz, Rhodes & Hanson, 1981; McDonnell & Wyn Jones, 1988). Choline used for betaine synthesis is derived from this phosphatidylcholine, and is associated with membrane turnover (Giddings & Hanson, 1982). In the chenopods (spinach and sugar beet) the pathway is entirely soluble and involves only the phosphorylated intermediates (Coughlan & Wyn Jones, 1982; Hanson & Rhodes, 1983; Weretilnyk & Summers, 1992). The origin of choline in other families has not been established.

The pathway from choline to glycinebetaine (Fig. 4) is common to all plants and consists of the two-stage oxidation of choline via betaine aldehyde (Lerma, Hanson & Rhodes, 1988; Lerma *et al.*, 1991; Weretilnyk *et al.*, 1989). The first step is catalysed by a ferredoxin-dependent choline monooxygenase (CMO) located in the chloroplast stroma (Brouquisse *et al.*, 1989; Lerma, Hanson & Rhodes, 1988). CMO has a molecular weight of 98 kDa, a pH optimum of 8, and is stimulated by Mg^{2+} (Rhodes & Hanson, 1993).

The second step, the oxidation of betaine aldehyde, is catalysed by an NAD^+-dehydrogenase (BADH) which is also mainly located in the chloroplast stroma, at least in spinach (Hanson *et al.*, 1985; Weigel, Lerma & Hanson, 1988; Weigel, Weretilnyk & Hanson, 1986; Weretilnyk & Hanson, 1989, 1990). BADH is a dimer of subunits with a molecular weight of about 60 kDa, has a pH optimum of 8.5 and is specific for betaine aldehyde (Arakawa *et al.*, 1987; Pan *et al.*, 1981; Weigel *et al.*, 1986; Weretilnyk & Hanson, 1988, 1989). BADH levels, expression and properties have been studied in higher plants and bacteria (Arakawa, Mizuno & Takabe, 1992*a,b*; Arakawa, Katayama & Takabe, 1990; Arakawa, *et al.*, 1987; Falkenberg & Strom, 1990; Ishitani *et al.*, 1993; Liang *et al.*, 1991; McCue & Hanson, 1992*a,b*; Pan, 1983,1988; Pan *et al.*, 1981; Weretilnyk & Hanson, 1988,1989). McCue and Hanson (1992*a*) implicated a non-hydraulic signal, which was not NaCl concentration, in the induction of BADH transcription in sugar beet in response to salt stress. cDNA BADH clones have been obtained from a number of higher plants (Ishitani & Takabe, 1992; McCue & Hanson, 1992*a,b*; Rhodes & Hanson, 1993; Weretilnyk & Hanson, 1990; Wood & Goldsbrough, 1993) and show good similarity (except for the transport peptide) with bacterial BADH genes (Andresen *et al.*, 1988; Boyd *et al.*, 1991; Lamark *et al.*, 1991).

Glycinebetaine synthesis is closely associated with chloroplasts which supply not only the basic carbon skeleton (in the form of glycine and

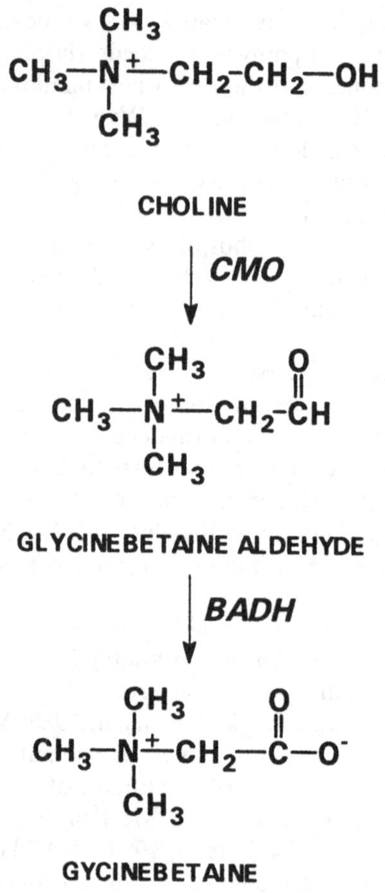

$$CH_3-\overset{\overset{\displaystyle CH_3}{|}}{\underset{\underset{\displaystyle CH_3}{|}}{N}}{}^{\pm}-CH_2-CH_2-OH$$

CHOLINE

↓ *CMO*

$$CH_3-\overset{\overset{\displaystyle CH_3}{|}}{\underset{\underset{\displaystyle CH_3}{|}}{N}}{}^{\pm}-CH_2-\overset{\overset{\displaystyle O}{\|}}{C}H$$

GLYCINEBETAINE ALDEHYDE

↓ *BADH*

$$CH_3-\overset{\overset{\displaystyle CH_3}{|}}{\underset{\underset{\displaystyle CH_3}{|}}{N}}{}^{\pm}-CH_2-\overset{\overset{\displaystyle O}{\|}}{C}-O^-$$

GYCINEBETAINE

Fig. 4. The two-stage oxidation of choline to betaine.

serine produced druing photorespiration), but also reducing power and methyl donors. In animals glycinebetaine can serve as a methyl donor, and in bacteria it can be catabolised. In plants there is little evidence of further metabolism of glycinebetaine, and some of the betaine is exported in the phloem to sink tissues (McDonnell & Wyn Jones, 1988; Hanson & Wyse, 1982).

Other betaines

For β-alaninebetaine and prolinebetaine radiotracer evidence suggests the direct methylation of the corresponding amino acids (Essery, McCaldin & Marion, 1962; Hanson *et al.*, 1991; Larher, 1976; Naidu *et al.*, 1987). DMSP is derived from methionine via S-methylmethionine (Greene, 1962; Rhodes & Hanson, 1993). A choline sulphotransferase produces choline-O-sulphate from choline in the Plumbaginaceae (Hanson *et al.*, 1991) and in bacteria (Fitzgerald & Luschinski, 1977).

Future prospects

Since the unique pathway from choline to glycinebetaine involves only two steps, it is a potential subject of transformation experiments. Betaine aldehyde dehydrogenase from the bacterium *E. coli* and the higher plant *Kochia scoparia* has already been expressed in transgenic tobacco (Jain, Jana & Selvaraj, 1993), although work on the choline monooxygenase is needed to complete the pathway. Since most plants produce at least small quantities of glycinebetaine it may also be possible to manipulate expression of existing genes. The advantages of increasing glycinebetaine accumulation in non-accumulating crop plants (e.g. rice, Rathinasabapathi *et al.*, 1993) subjected to stress are evident, but there are also other possibilities. One is the uncoupling of the choline pathway in cereals from membrane turnover by introducing the phosphorylcholine to choline step from chenopods. There are also reasons for wanting to inhibit glycinebetaine accumulation in some plants by, for example, introducing anti-sense genes. These include production of betaine-deficient lines for physiological experiments and prevention of nitrogen being 'locked up' in non-metabolized betaines. Betaine is also an undesirable impurity in sugar beet, interfering with sugar production. Finally, there is the suggestion that glycinebetaine promotes the growth of fungal pathogens (Pearce, Strange & Smith, 1976) and aphids (Corcuera, 1993; Salas & Corcuera, 1991; Zuniga & Corcuera, 1987*a,b*). Manipulation of other betaine pathways may also be of interest. β-Alaninebetaine has been suggested to be more useful than glcyinebetaine in plants growing in hypoxic conditions since methylaltion of β-alanine does not require oxygen. Choline-O-sulphate or DMSP production might be beneficial in high sulphur environments. Another interesting possibility is the accumulation of the more stable prolinebetaine instead of, or as well as, the labile compatible solute, proline.

References

Ahmad, N. & Wyn Jones, R.G. (1979). Glycinebetaine, proline and inorganic ion levels in barley seedlings following transient stress. *Plant Science Letters*, **15**, 231–7.

Ahmad, N., Wyn Jones, R.G. & Jeschke, W.D. (1987). Effect of exogenous glycinebetaine on Na⁺ transport in barley roots. *Journal of Experimental Botany*, **38**, 913–22.

Andresen, P.A., Kaasen, I., Styrvold, O.B., Boulnois, G. & Strom, A.R. (1988). Molecular-cloning, physical mapping and expression of the bet genes governing the osmoregulatory choline glycine betaine pathway of *Escherichia coli*. *Journal of General Microbiology*, **134**, 1737–46.

Arakawa, K., Mizuno, K. & Takabe, T. (1992*a*). Betaine aldehyde dehydrogenase protein is present in leaves of both betaine accumulators and nonaccumulators in various cereal plants. *Photosynthesis Research*, **34**, 218.

Arakawa, K, Mizuno, K. & Takabe T. (1992*b*). Immunological studies of betaine aldehyde dehydrogenase in barley. *Photosynthesis Research*, **34**, 217.

Arakawa, K., Katayama, M. & Takabe, T. (1990). Levels of betaine and betaine aldehyde dehydrogenase–activity in the green leaves, and etiolated leaves and roots of barley. *Plant and Cell Physiology*, **31**, 797–803.

Arakawa, K., Takabe, T., Sugiyama, T. & Akazawa, T. (1987). Purification of betaine-aldehyde dehydrogenase from spinach leaves and preparation of its antibody. *Journal of Biochemistry*, **101**, 1485–8.

Araya, F., Abarca, O., Zuniga, G.E., Corcuera, L.J. (1991). Effects of NaCl on glycine-betaine and on aphids in cereal seedlings. *Phytochemistry*, **30**, 1793–5.

Barak, A.J. & Tuma, D.J. (1979). A simplified procedure for the determination of betaine in liver. *Lipids*, **14**, 860–3.

Bercht, C.A.L., Lousberg, R.J.J.Ch., Küppers, F.J.E.M. & Salemink, C.A. (1973). L-(+)-Isoleucine betaine in *Cannabis* seeds. *Phytochemistry*, **12**, 2457–9.

Bergmann, H. & Eckert, H. (1984). Effect of glycinebetaine on water-use efficiency of wheat (*Triticum-aestivum* L). *Biologia Plantarum*, **26**, 384–7.

Bernard, T., Goas, G., Hamelin, J. & Joucla, M. (1981). Characterization of DOPA-betaine, tyrosine-betaine and N-dimethyltyrosine from *Lobaria laetevirens*. *Phytochemistry*, **20**, 2325–6.

Billard, J.P., Rouxel, M.F. & Boucaud, J. (1986). Implication of glycine betaine in the control *in situ* of ribonuclease-activity in excised leaves of *Suaeda maritima* (L) Dum. var. *macrocarpa* Moq. *Comptes Rendus de l Academie des Sciences Serie III-Sciences de la Vie*, **303**, 291–4.

Blunden, G., El Barouni, M., Gordon, S.M., McLean, W.F.H. & Rogers, D.J. (1981). Extraction, purification and characterisation of Dragendorff-positive compounds from some British marine algae. *Botanica Marina*, **24**, 451–6.

Blunden, G. & Gordon, S.M. (1986). Betaines and their sulphonio analogues in marine algae. *Progress in Phycological Research*, **4**, 39–80.

Blunden, G., Gordon, S.M., Crabb, T.A., Roch, O.G., Rowan, M.G. & Wood, B. (1986). NMR spectra of betaines from marine algae. *Magnetic Resonance in Chemistry*, **24**, 965–71.

Blunden, G., Smith, B.E., Irons, M.W., Yang, M.H., Roch, O.G. & Patel, A.V. (1992). Betaines and tertiary sulfonium compounds from 62 species of marine-algae. *Biochemical Systematics and Ecology*, **20**, 4, 373–88.

Bouillard, L. & LeRudulier, D. (1983). Nitrogen-fixation under osmotic-stress – enhancement of nitrogenase biosynthesis in *Klebsiella pneumoniae* by glycine betaine. *Physiologie Vegetale*, **21**, 447–57.

Boyd, L.A., Adam, L., Pelcher, L.E., McHughen, A., Hirji, R. & Selvaraj, G. (1991). Characterization of an *Escherichia coli* gene encoding betaine aldehyde dehydrogenase (BADH) – Structural similarity to mammalian ALDHS and a plant BADH. *Gene*, **103**, 45–52.

Brady, C.J., Gibson, T.S., Barlow, E.W.R., Speirs, J. & Wyn Jones, R.G. (1984). Salt-tolerance in plants. I. Ions, compatible organic solutes and the stability of plant ribosomes. *Plant Cell and Environment*, **7**, 571–8.

Brouquisse, R., Weigel, P., Rhodes, D., Yocum, C.F. & Hanson, A.D. (1989). Evidence for a ferredoxin-dependent choline monooxygenase from spinach chloroplast stroma. *Plant Physiology*, **90**, 322–9.

Brunk, D.G., Rich, P.J. & Rhodes, D. (1989). Genotypic variation for glycinebetaine among public inbreds of maize. *Plant Physiology*, **91**, 1122–5.

Cavalieri, A.J. (1983). Proline and glycinebetaine accumulation by *Spartina alterniflora* Loisel in response to NaCl and nitrogen in a controlled environment. *Oecologia*, **57**, 20–4.

Cavalieri, A.J. & Huang, A.H.C. (1981). Accumulation of proline and glycinebetaine in *Spartina alterniflora* Loisel in response to NaCl and nitrogen in the marsh. *Oecologia*, **49**, 224–8.

Chambers, S.T., Kunin, C.M., Miller, D. & Hamada, A. (1987). Dimethylthetin can substitute for glycine betaine as an osmoprotectant molecule for *Escherichia coli*. *Journal of Bacteriology*, **169**, 4845–7.

Chastellain, F. & Hirsbrunner, P. (1976). Determination of betaine and choline in feeds by ^{1}H-NMR spectrometry. *Zeitschrift für Analytische Chemie*, **278**, 207–8.

Corcuera, L.J. (1993). Biochemical basis for the resistance of barley to aphids. *Phytochemistry*, **33**, 741–7.

Cornforth J.W. & Henry A.J. (1952). The isolation of L-stachydrine from the fruit of *Capparis tomentosa*. *Journal of the Chemical Society*, 601–3.

Coughlan, S.J. & Heber, U. (1982). The role of glycinebetaine in the protection of spinach thylakoids against freezing stress. *Planta*, **156**, 62–9.

Coughlan, S.J. & Wyn Jones, R.G. (1982). Glycinebetaine biosynthesis and its control in detached secondary leaves of spinach. *Planta*, **154**, 6–17.

Cromwell, B.T. & Rennie, S.D. (1954). The biosynthesis and metabolism of betaines in plants. I. The estimation and distribution of glycinebetaine in *Beta vulgaris*. L. and other plants. *Biochemical Journal*, **55**, 189–92.

Csonka, L.N. (1989). Physiological and genetic responses of bacteria to osmotic stress. *Microbiological Reviews*, **53**, 121–47.

Csonka, L.N. & Hanson, A.D. (1991). Prokaryotic osmoregulation: genetics and physiology. *Annual Reviews of Microbiology*, **45**, 569–606.

Dacey, J.W.H., King, G.M. & Wakeham, S.G. (1987). Factors controlling emission of dimethylsulfide from salt marshes. *Nature, London*, **330**, 643–5.

Delaveau, P., Koudogbo, B. & Pousset, J.L. (1973). Alkaloids in Capparidaceae. *Phytochemistry* **12**, 2893–5.

Delwiche C.C. & Bergoff, H.M. (1957). Pathway of betaine and choline synthesis in *Beta vulgaris*. *Journal of Biological Chemistry*, **233**, 430–3.

Dickson, D.M.J. & Kirst, G.O. (1986). The role of beta-dimethylsulphoniopropionate, glycine betaine and homarine in the osmoacclimation of *Platymonas subcordiformis*. *Planta*, **167**, 536–43.

Duke, E.R., Johnson, C.R. & Koch, K.E. (1986). Accumulation of phosphorus, dry-matter and betaine during NaCl stress of split-root citrus seedlings colonized with vesicular arbuscular mycorrhizal fungi on zero, one or two halves. *New Phytologist*, **104**, 583–90.

Essery, J.M., McCaldin, D.J. & Marion, L. (1962). The biogenesis of stachydrine. *Phytochemistry*, **1**, 209–13.

Falkenberg, P. & Strom, A.R. (1990). Purification and characterization of osmoregulatory betaine aldehyde dehydrogenase of *Escherichia coli*. *Biochimica et Biophysica Acta*, **1034**, 253–9.

Fitzgerald, J.W. & Luschinski, P.C. (1977). Further studies on the formation of choline sulfate by bacteria. *Canadian Journal of Microbiology*, **23**, 483–90.

Ford, C.W. & Wilson, J.R. (1981). Changes in levels of solutes during osmotic adjustment to water stress in leaves of four tropical pasture species. *Australian Journal of Plant Physiology*, **8**, 77–91.

Fougere, F. & LeRudulier, D. (1990a). Glycine betaine biosynthesis and catabolism in bacteroids of *Rhizobium meliloti* – effect of salt stress. *Journal of General Microbiology*, **136**, 2503–10.

Fougere, F. & LeRudulier, D. (1990b). Uptake of glycine betaine and its analogs by bacteroids of *Rhizobium meliloti*. *Journal of General Microbiology*, **136**, 157–63.

Gamboa, A., Valenzuela, E.M. & Murillo, E. (1991). Biochemical changes due to water-loss in leaves of *Amaranthus hypochondriacus* L. *Journal of Plant Physiology*, **137**, 586–90.

Génard, H., Le Saos, J., Billard, J.P., Trémolières, A. & Boucaud, J. (1991). Effect of salinity on lipid-composition, glycine betaine content and photo-synthetic activity in chloroplasts of *Suaeda maritima*. *Plant Physiology and Biochemistry*, **29**, 421–7.

Gibson, T.S., Speirs, J. & Brady, C.J. (1984). Salt-tolerance in plants. II. In vitro translation of m-RNAs from salt-tolerant and salt-sensitive plants on wheat germ ribosomes. Responses to ions and compatible organic solutes. *Plant Cell and Environment*, **7**, 579–87.

Giddings, T.H. & Hanson, A.D. (1982). Water stress provokes a generalized increase in phosphatidylcholine turnover in barley leaves. *Planta*, **155**, 493–501.

Gloux, K. & LeRudulier, D. (1988). Osmoregulation in *Rhizobium meliloti* – role of proline betaine. *Plant Physiology and Biochemistry*, **26**, 229.

Goas, G., Goas, M. & Larher, F. (1982). Accumulation of free proline and glycine betaine in *Aster tripolium* subjected to a saline shock – a kinetic study related to light period. *Physiologia Plantarum*, **55**, 383–8.

Gorham, J. (1984). Separation of plant betaines and their sulphur analogues by cation-exchange high-performance liquid chromatography. *Journal of Chromatography*, **287**, 345–51.

Gorham, J. (1986). Separation and quantitative estimation of betaine esters by high-performance liquid-chromatography. *Journal of Chromatography*, **361**, 301–10.

Gorham, J., Coughlan, S.J., Storey, R. & Wyn Jones, R.G. (1981). Estimation of quaternary ammonium and tertiary sulphonium compounds by thin-layer electrophoresis and scanning reflectance densitometry. *Journal of Chromatography*, **210**, 550–4.

Gorham, J., Hughes, Ll. & Wyn Jones, R.G. (1980). Chemical composition of salt marsh plants from Ynys Môn (Anglesey): the concept of physiotypes. *Plant Cell and Environment*, **3**, 309–18.

Gorham, J. & McDonnell, E. (1985). High-performance liquid-chromatographic method for the separation and estimation of choline, glycinebetaine aldehyde and related-compounds. *Journal of Chromatography*, **350**, 245–54.

Gorham, J., McDonnell, E. & Wyn Jones, R.G. (1982). Determination of betaines as ultraviolet-absorbing esters. *Analytica Chimica Acta*, **138**, 277–83.

Grattan, S.R. & Grieve, C.M. (1985). Betaine status in wheat in relation to nitrogen stress and to transient salinity stress. *Plant and Soil*, **85**, 3–9.

Greene, R.C. (1962). Biosynthesis of dimethyl-β-propiothetin. *Journal of Biological Chemistry*, **237**, 2251–4.

Grieve, C.M. & Grattan, S.R. (1983). Rapid assay for determination of water soluble quaternary ammonium compounds. *Plant and Soil*, **70**, 303–7.

Grieve, C.M. & Maas, E.V. (1984). Betaine accumulation in salt-stressed sorghum. *Physiologia Plantarum*, **61**, 167–71.

Gröne, T. & Kirst, G.O. (1991). Aspects of dimethylsulfoniopropionate effects on enzymes isolated from the marine phytoplankter *Tetraselmis subscordiformis* (Stein). *Journal of Plant Physiology*, **138**, 85–91.

Grumet, R., Albrechtsen, R.S. & Hanson, A.D. (1987). Growth and yield of barley isopopulations differing in solute potential. *Crop Science*, **27**, 991–5.

Grumet, R. & Hanson, A.D. (1983). Genetic analyses of glycinebetaine accumulation in barley. *Hortscience*, **18**, 557.

Grumet, R. & Hanson, A.D. (1986). Genetic-evidence for an osmoregulatory function of glycinebetaine accumulation in barley. *Australian Journal of Plant Physiology*, **13**, 353–64.

Grumet, R., Isleib, T.G. & Hanson, A.D. (1985). Genetic-control of glycinebetaine level in barley. *Crop Science*, **25**, 618–22.

Guy, R.D., Warne, P.G. & Reid, D.M. (1984). Glycinebetaine content of halophytes – improved analysis by liquid-chromatography and interpretations of results. *Physiologia Plantarum*, **61**, 195–202.

Hall, J.L., Harvey, D.M.R. & Flowers, T.J. (1978). Evidence for the cytoplasmic localization of betaine in leaf cells of *Suaeda maritima*. *Planta*, **140**, 59–62.

Hanson, A.D. & Gage, D.A. (1991). Identification and determination by fast-atom-bombardment mass-spectrometry of the compatible solute choline-O-sulfate in *Limonium* species and other halophytes. *Australian Journal of Plant Physiology*, **18**, 317–27.

Hanson, A.D., Huang, Z.-H. & Gage, D. (1993). Evidence that the putative compatible solute 5-dimethylsulfoniopentanoate is an extraction artifact. *Plant Physiology*, **101**, 1391–3.

Hanson, A.D., May, A.M., Grumet, R., Bode, J., Jamieson, G.C. & Rhodes, D. (1985). Betaine synthesis in chenopods – localization in chloroplasts. *Proceedings of the National Academy of Sciences, USA*, **82**, 3678–82.

Hanson, A.D. & Nelsen, C.E. (1978). Betaine accumulation and [^{14}C]formate metabolism in water-stressed barley leaves. *Plant Physiology*, **62**, 305–12.

Hanson, A.D., Rathinasabapathi, B., Chamberlin, B. & Gage, D.A. (1991). Comparative physiological evidence that beta-alanine betaine and choline-O-sulfate act as compatible osmolytes in halophytic *Limonium* species. *Plant Physiology*, **97**, 1199–205.

Hanson, A.D. & Rhodes, D. (1983). C-14 tracer evidence for synthesis of choline and betaine via phosphoryl base intermediates in salinized sugarbeet leaves. *Plant Physiology*, **71**, 692–700.

Hanson, A.D. & Scott, N.A. (1980). Betaine synthesis from radioactive precursors in attached, water-stressed barley leaves. *Plant Physiology*, **66**, 342–8.

Hanson, A.D. & Wyse, R. (1982). Biosynthesis, translocation, and accumulation of betaine in sugar-beet and its progenitors in relation to salinity. *Plant Physiology*, **70**, 1191–8.

Higgins, J., Hodges, N.A., Olliff, C.J. & Phillips, A.J. (1987). A comparative investigation of glycinebetaine and dimethylsulfoxide as liposome cryoprotectants. *Journal of Pharmacy and Pharmacology*, **39**, 577–82.

Hitz, W.D. & Hanson, A.D. (1980). Determination of glycine betaine by pyrolysis-gas chromatography in cereals and grasses. *Phytochemistry*, **19**, 2371–4.

Hitz, W.D., Ladyman, J.A.R. & Hanson, A.D. (1982). Betaine synthesis and accumulation in barley during field water-stress. *Crop Science*, **22**, 47–54.

Hitz, W.D., Rhodes, D. & Hanson, A.D. (1981). Radiotracer evidence implicating phosphoryl and phosphatidyl bases as intermediates in betaine synthesis by water-stressed barley leaves. *Plant Physiology*, **68**, 814–22.

Huber, W. & Eder, A. (1982). Proline and glycinebetaine accumulation in leaves and shoots of *Euphorbia trigona* and their role during water deficiency. *Biochemie und Physiologie der Pflanzen*, **177**, 184–91.

Imhoff, J.F. & Rodriguez Valera, F. (1984). Betaine is the main compatible solute of halophilic eubacteria. *Journal of Bacteriology*, **160**, 478–9.

Incharoensakdi, A., Takabe, T. & Akazawa, T. (1986). Effect of betaine on enzyme-activity and subunit interaction of ribulose-1,5-bisphosphate carboxylase oxygenase from *Aphanothece halophytica*. *Plant Physiology*, **81**, 1044–9.

Ishitani, M. & Takabe, T. (1992). Molecular-cloning of betaine aldehyde dehydrogenase from barley and regulation of its expression by osmotic-stress. *Photosynthesis Research*, **34**, 217.

Ishitani, M., Arakawa, K., Mizuno, K., Kishitani, S. & Takabe, T. (1993). Betaine aldehyde dehydrogenase in the Gramineae – levels in leaves of both betaine-accumulating and nonaccumulating cereal plants. *Plant and Cell Physiology*, **34**, 493–5.

Itai, C. & Paleg, L.G. (1982). Responses of water-stressed *Hordeum distichum* L and *Cucumis sativus* to proline and betaine. *Plant Science Letters*, **25**, 329–35.

Jain, R.K., Jana, S. & Selvaraj, G. (1993). Genetic manipulation of betaine aldehyde oxidation in tobacco. *Plant Physiology*, **102**, 166.

Jolivet, Y., Hamelin, J. & Larher, F. (1983). Osmoregulation in halophytic higher-plants – the protective effects of glycine betaine and other related solutes against the oxalate destabilization of membranes in beet root-cells. *Zeitschrift für Pflanzenphysiologie*, **109**, 171–80.

Jolivet, Y., Larher, F. & Hamelin, J. (1982). Osmoregulation in halophytic higher-plants – the protective effect of glycine betaine against the heat destabilization of membranes. *Plant Science Letters*, **25**, 193–201.

Jones, G.P., Naidu, B.P., Paleg, L.G., Tiekink, E.R.T. & Snow, M.R. (1987). 4-hydroxy-N-methylproline analogues in *Melaleuca* spp. *Phytochemistry*, **26**, 3343–4.

Jones, G.P., Naidu, B.P., Starr, R.K. & Paleg, L.G. (1986). Estimates of solutes accumulating in plants by ^{1}H nuclear magnetic resonance spectroscopy. *Australian Journal of Plant Physiology*, **13**, 6449–658.

Koheil, M.A.H., Hilal, S.H., Elalfy, T.S. & Leistner, E. (1992). Quaternary ammonium-compounds in intact plants and cell-suspension cultures of *Atriplex semibaccata* and *A. halimus* during osmotic-stress. *Phytochemistry*, **31**, 2003–8.

Koskinen, E., Junnila, M., Katila, T. & Soini, H. (1989). A preliminary-study on the use of betaine as a cryoprotective agent in deep freezing of stallion semen. *Journal of Veterinary Medicine Series A-Animal Physiology Pathology and Clinical Veterinary medicine – Zentralblatt fur Veterinarmedizin Reihe A*, **36** 110–14.

Ladyman, J.A.R., Ditz, K.M., Grumet, R. & Hanson, A.D. (1983). Genotypic variation for glycinebetaine accumulation by cultivated and wild barley in relation to water-stress. *Crop Science*, **23**, 465–8.

Ladyman, J.A.R., Hitz, W.D. & Hanson, A.D. (1980). Translocation and metabolism of glycine betaine by barley plants in relation to water stress. *Planta*, **150**, 191–6.

Lamark, T., Kaasen, I., Eshoo, M.W., Falkenberg, P., McDougall, J. & Strom, A.R. (1991). DNA-sequence and analysis of the BET genes encoding the osmoregulatory choline glycine betaine pathway of *Escherichia coli*. *Molecular Microbiology*, **5**, 1049–64.

Larher, F. (1976). Sur quelques particularités du métabolisme azoté d'une halophyte: Limonium vulgare Mill. DSc Thesis, Université de Rennes.

Larher, F. (1988). Natural abundance ^{13}C-nuclear magnetic resonance studies on the compatible solutes of halophytic higher plants. *Plant Physiology and Biochemistry*, **26**, 35–45.

Larher, F. & Hamelin, J. (1979). L'acide diméthylsulfonium-5 penta-noïque de *Diplotaxis tenuifolia*. *Phytochemistry*, **18**, 1396–7.

Larher, F., Hamelin, J. & Stewart, G.R. (1977). L'acide diméthylsul-fonium-3 propanoïque de *Spartina anglica*. *Phytochemistry*, **16**, 2019–20.

Larher, F., Jolivet, Y., Briens, M. & Goas, M. (1982). Osmoregulation in higher-plant halophytes – organic nitrogen accumulation in glycine betaine and proline during the growth of *Aster tripolium* and *Suaeda macrocarpa* under saline conditions. *Plant Science Letters*, **24**, 201–10.

Laurie, S. & Stewart, G.R. (1990). The effects of compatible solutes on the heat stability of glutamate synthase from chickpeas grown under different nitrogen and temperature regimes. *Journal of Experimental Botany*, **41**, 1415–22.

Leigh, R.A., Ahmad, N. & Wyn Jones, R.G. (1981). Assessment of glycinebetaine and proline compartmentation by analysis of isolated beet vacuoles. *Planta*, **153**, 34–41.

Lerma, C., Hanson, A.D. & Rhodes, D. (1988). Oxygen–18 and deuterium labeling studies of choline oxidation by spinach and sugar beet. *Plant Physiology*, **88**, 695–702.

Lerma, C., Rich, P.J., Ju, G.C., Yang, W.J., Hanson, A.D. & Rhodes, D. (1991). Betaine deficiency in maize – complementation tests and metabolic basis. *Plant Physiology*, **95**, 1113–19.

Le Rudulier, D., Bernard, T., Goas, G. & Hamelin, J. (1984a). Osmoregulation in *Klebsiella pneumoniae* – enhancement of anaerobic growth and nitrogen-fixation under stress by proline betaine, gamma-butyrobetaine, and other related compounds. *Canadian Journal of Microbiology*, **30**, 299–305.

Le Rudulier, D., Bernard, T., Pocard, J.A. & Goas, G. (1983). Enhancement of osmotolerance in *Rhizobium meliloti* by glycinebetaine and proline betaine. *Comptes Rendus de l' Academie des Sciences Serie III – Sciences de la Vie*, **297**, 155.

Le Rudulier, D., Strøm, A.R., Dandekar, A.M., Smith, L.T. & Valentine, R.C. (1984b). Molecular biology of osmoregulation. *Science*, **224**, 1064–8.

Liang, Z., Zhao, Y., Li, Y.C., Zou, Y.P., Luo, A.L. & Huang, J.F. (1991). Inhibitory factor of betaine aldehyde dehydrogenase from spinach leaves. *Chinese Science Bulletin*, **36**, 1477–80.

Lloyd, A.W., Baker, J.A., Smith, G., Olliff, C.J. & Rutt, K.J. (1992). A comparison of glycine, sarcosine, N,N-dimethylglycine, glycinebetaine and N-modified betaines as liposome cryoprotectants. *Journal of Pharmacy and Pharmacology*, **44**, 507–11.

Lone, M.I., Kueh, J.S.H., Jones, R.G.W. & Bright, S.W.J. (1987). Influence of proline and glycinebetaine on salt tolerance of cultured barley embryos. *Journal of Experimental Botany*, **38**, 479–90.

Low, P.S. (1985). Molecular basis of the biological compatibility of nature's osmolytes. In *Transport Processes, Iono- and Osmoregulation* ed. R. Gilles, M. Gilles-Baillien. pp. 469–77. Berlin, Springer.

Manetas, Y. (1990). A re-examination of NaCl effects on phosphoenol pyruvate carboxylase at high (physiological) enzyme concentrations. *Physiologia Plantarum*, **78**, 225–9.

Matoh, T., Watanabe, J. & Takahashi, E. (1987). Sodium, potassium, chloride, and betaine concentrations in isolated vacuoles from salt-grown *Atriplex gmelini* leaves. *Plant Physiology*, **84**, 173–7.

McCue, K.F. & Hanson, A.D. (1992a). Effects of soil-salinity on the expression of betaine aldehyde dehydrogenase in leaves – investigation of hydraulic, ionic and biochemical signals. *Australian Journal of Plant Physiology*, **19**, 555–64.

McCue, K.F. & Hanson, A.D. (1992b). Salt-inducible betaine aldehyde dehydrogenase from sugar-beet – cDNA cloning and expression. *Plant Molecular Biology*, **18**, 1–11.

McDonnell, E. & Wyn Jones, R.G. (1988). Glycinebetaine biosynthesis and accumulation in unstressed and salt-stressed wheat. *Journal of Experimental Botany*, **39**, 421–30.

McLean, W.F.H. (1976). Studies in the family Capparidaceae with special reference to the genus *Maerua*. PhD Thesis, CNAA.

Minkler, P.E., Ingalls, S.T., Kormos, L.S., Weir, D.E. & Hoppel, C.L. (1984). Determination of carnitine, butyrobetaine, and betaine as 4'-bromophenacyl ester derivatives by high-performance liquid-chromatography. *Journal of Chromatography*, **336**, 271–83.

Monyo, E.S., Ejeta, G. & Rhodes, D. (1992). Genotypic variation for glycinebetaine in sorghum and its relationship to agronomic and morphological traits. *Maydica*, **37**, 283–6.

Mudd, S.H. & Datko, A.H. (1989). Synthesis of ethanolamine and its regulation in *Lemna paucicostata*. *Plant Physiology*, **91**, 587–97.

Murata, N., Mohanty, P.S., Hayashi, H. & Papageorgiou, G.C. (1992).Glycinebetaine stabilizes the association of extrinsic proteins with the photosynthetic oxygen-evolving complex. *FEBS Letters*, **296**, 187–9.

Murray, A.J.S. & Ayres, P.G. (1986). Infection with powdery mildew can enhance the accumulation of proline and glycinebetaine by salt stressed barley seedlings. *Physiological and Molecular Plant Pathology*, **29**, 271–7.

Naidu, B.P., Jones, G.P., Paleg, L.G. & Poljakoff-Mayber, A. (1987). Proline analogues in *Melaleuca* species: response of *Melaleuca lanceolata* and *M. uncinata* to water stress and salinity. *Australian Journal of Plant Physiology*, **14**, 669–77.

Naidu, B.P., Paleg, L.G., Aspinall, D., Jennings, A.C. & Jones, G.P. (1990). Rate of imposition of water stress alters the accumulation of nitrogen-containing solutes by wheat seedlings. *Australian Journal of Plant Physiology*, **17**, 653–64.

Naidu, B.P., Paleg, L.G., Aspinall, D., Jennings, A.C. & Jones, G.P. (1991). Amino-acid and glycine betaine accumulation in cold-stressed wheat seedlings. *Phytochemistry*, **30**, 407–9.

Naidu, B.P., Paleg, L.G. & Jones, G.P. (1992). Nitrogenous compatible solutes in drought-stressed *Medicago* spp. *Phytochemistry*, **31**, 1195–7.

Nash, D., Paleg, L.G. & Wiskich, J.T. (1982). Effect of proline, betaine and some other solutes on the heat-stability of mitochondrial-enzymes. *Australian Journal of Plant Physiology*, **9**, 47–57.

Nikolopoulus, D. & Manetas, Y. (1991). Compatible solutes and in vitro stability of *Salsola soda* enzymes: proline incompatibility. *Phytochemistry*, **30**, 411–3.

Okogun, J.I., Spiff, A.I. & Ekong, D.E.U. (1976). Triterpenoids and betaines from the latex and bark of *Antiaris africana*. *Phytochemistry*, **15**, 826–7.

Paleg, L.G., Douglas, T.J., van Daal, A. & Keech, D.B. (1981). Proline, betaine and other organic solutes protect enzymes against heat inactivation. *Australian Journal of Plant Physiology*, **8**, 107–14.

Paleg, L.G., Stewart, G.R. & Bradbeer, J.W. (1984). Proline and glycine betaine influence protein solvation. *Plant Physiology*, **75**, 974–8.

Pan, S.M. (1983). The effect of salt stress on the betaine aldehyde dehydrogenase in spinach. *Taiwania*, **28**, 128–37.

Pan, S.M. (1988). Betaine aldehyde dehydrogenase in spinach. *Botanical Bulletin of Academia Sinica*, **29**, 255–63.

Pan, S.M., Moreau, R.A., Yu, C. & Huang, A.H.C. (1981). Betaine accumulation and betaine-aldehyde dehydrogenase in spinach leaves. *Plant Physiology*, **67**, 1105–8.

Papageorgiou, G.C., Fujimura, Y. & Murata, N. (1991). Protection of the oxygen-evolving photosystem-II complex by glycinebetaine. *Biochimica et Biophysica Acta*, **1057**, 361–6.

Paquet, L., Rathinasabapathi, B., Saini, H., Zamir, L., Gage, D.A., Huang, T.H. and Hanson, A.D. (1994) Accumulation of the compatible solute 3-dimethylsulfoniopropionate in sugarcane and its relatives, but not other graminaceous crops. *Australian Journal of Plant Physiology*, **21**, 37–48.

Pearce, R.B., Strange, R.N. & Smith, H. (1976). Glycinebetaine and choline in wheat: Distribution and relation to infection by *Fusarium graminearium*. *Phytochemistry*, **15**, 953–4.

Pocard, J.A., Bernard, T., Gas, G. & Le Rudulier, D. (1984). Partial restoration of nitrogen-fixation activity, by glycine betaine and proline betaine, in young *Medicago sativa* L. plants under water-stress. *Comptes Rendus de l Academie des Sciences Serie III-Sciences de la Vie*, **298**, 477–80.

Pocard, J.A., Bernard, T. & Le Rudulier, D. (1991). Translocation and metabolism of glycine betaine in nodulated alfalfa plants subjected to salt stress. *Physiologia Plantarum*, **81**, 95–102.

Pocard, J.A. & Smith, L.T. (1988). Glycine betaine metabolism in *Rhizobium meliloti* – osmotic control and activation by choline. *Plant Physiology and Biochemistry*, **26**, 224.

Pollard, A. & Wyn Jones, R.G. (1979). Enzyme activities in concentrated solutions of glycinebetaine and other solutes. *Planta*, **144**, 291–8.

Rajakylä, E. & Paloposki, M. (1983). Determination of sugars (and betaine) in molasses by high-performance liquid chromatography. Comparison of the results with those obtained by the classical Lane–Eynon method. *Journal of Chromatography*, **282**, 595–602.

Rathinasabapathi, B., Gage, D.A., MacKill, D.J. & Hanson, A.D. (1993). Cultivated and wild rices do not accumulate glycinebetaine due to deficiencies in 2 biosynthetic steps. *Crop Science*, **33**, 534–8.

Rhodes, D. & Hanson, A.D. (1993). Quaternary ammonium and tertiary sulfonium compounds in higher-plants. Annual Review of Plant *Physiology and Plant Molecular Biology*, **44**, 357–84.

Rhodes, D. & Rich, P.J. (1988). Preliminary genetic-studies of the phenotype of betaine deficiency in *Zea mays* L. *Plant Physiology*, **88**, 102–8.

Rhodes, D., Rich, P.J., Brunk, D.G., Ju, G.C., Rhodes, J.C., Pauly, M.H. & Hansen, L.A. (1989). Development of 2 isogenic sweet corn hybrids differing for glycinebetaine content. *Plant Physiology*, **91**, 1112–21.

Rhodes, D., Rich, P.J., Myers, A.C., Reuter, C.C. & Jamieson, G.C. (1987). Determination of betaines by fast atom bombardment mass-spectrometry – identification of glycine betaine deficient genotypes of *Zea mays*. *Plant Physiology*, **84**, 781–8.

Rhodes, D., Yang, W.J., Samaras, Y., Wood, K.V., Bonham, C.C., Rhodes, J.C. & Burr, B. (1993). Map locations of genes conferring glycinebetaine and trigonelline accumulation in maize. *Plant Physiology*, **102**, 160.

Robinson, S.P. & Jones, G.P. (1986). Accumulation of glycinebetaine in chloroplasts provides osmotic adjustment during salt stress. *Australian Journal of Plant Physiology*, **13**, 659–68.

Rudolph, A.S., Crowe, J.H. & Crowe, L.M. (1986). Effects of 3 stabilizing agents proline, betaine, and trehalose on membrane phospholipids. *Archives of Biochemistry and Biophysics*, **245**, 134–43.

Salas, M.L. & Corcuera, L.J. (1991). Effect of environment on gramine content in barley leaves and susceptibility to the aphid *Schizaphis graminum*. *Phytochemistry*, **30**, 3237–40.

Sato, N. (1992). Betaine lipids. *Botanical Magazine – Tokyo*, **105**, 185–97.

Sauvage, D., Hamelin, J. & Larher, F. (1983). Glycine betaine and other structurally related-compounds improve the salt tolerance of *Rhizobium meliloti*. *Plant Science Letters*, **31**, 291–302.

Schröppel-Meier, G. & Kaiser, W.M. (1988). Ion homeostasis in chloroplasts under salinity and mineral deficiency. I. Solute concentrations in leaves and chloroplasts from spinach plants under NaCl or NaNO$_3$ salinity. *Plant Physiology*, **87**, 822–7.

Selinioti, E., Nikolopoulus, D. & Manetas, Y. (1987). Organic co-solutes as stablilizers of phosphoenolpyruvate carboxylase in storage: an interpretation of their action. *Australian Journal of Plant Physiology*, **14**, 203–10.

Shomer Ilan, A., Jones, G.P. & Paleg, L.G. (1991). *In vitro* thermal and salt stability of pyruvate-kinase are increased by proline analogs and trigonelline. *Australian Journal of Plant Physiology*, **18**, 279–86.

Shrestha, T. & Bisset, N.G. (1991). Quaternary nitrogen-compounds from south-American Moraceae. *Phytochemistry*, **30**, 3285–7.

Smirnoff, N. & Stewart, G.R. (1985). Stress metabolites and their role in coastal plants. *Vegetatio*, **62**, 273–8.

Smith, L.T., Pocard, J.A., Bernard, T. & LeRudulier, D. (1988). Osmotic control of glycine betaine biosynthesis and degradation in *Rhizobium meliloti*. *Journal of Bacteriology*, **170**, 3142–9.

Somero, G.N. (1986). Protons, osmolytes, and fitness of internal milieu for protein function. *American Journal of Physiology*, **251**, R197–R213.

Steinle, G. & Fischer, E. (1978). Hochdruckflüssigkeitschromatographische Methode zur Bestimmung von Betain in Zuckerfabriksprodukten. *Zuckerindustrie*, **103**, 129–31.

Strøm, A.R., Le Rudulier, D., Jakowec, M.W. , Bunnell, R.C. & Valentine, R.C. (1983). Osmoregulatory (Osm) genes and osmoprotective compounds. In *Genetic Engineering of Plants: An Agricultural Perspective*, ed. T. Kosuge, C.P. Meredith & A. Hollaender. pp. 39–59. New York: Plenum.

Storey, R., Gorham, J., Pitman, M.G., Hanson, A.D. & Gage, D. (1993). Response of *Melanthera biflora* to salinity and water stress. *Journal of Experimental Botany*, **44**, 1551–60.

Storey, R. & Wyn Jones, R.G. (1977). Quaternary ammonium comounds in plants in relation to salt resistance. *Phytochemistry*, **16**, 447–53.

Takhtajan, A. (1969). *Flowering Plants – Origin and Dispersal*. Washington, Smithsonian Institute Press.

Tamada. M., Endo, K. & Hikino, H. (1978). Maokonine, hypertensive principle of *Ephedra* roots. *Planta Medica*, **34**, 291–3.

van Diggelen, J., Rozema, J. Dickson, D.M.J. & Broekman, R. (1986). β-3-Dimethylsulphoniopropionate, proline and quaternary ammonium compounds in *Spartina anglica* in relation to sodium chloride, nitrogen and sulphur. *New Phytologist*, **103**, 573–86.

Vialle, J., Kolosky, M. & Rocca, J.L. (1981). Determination of betaine in sugar and wine by liquid-chromatography. *Journal of Chromatography*, **204**, 429–35.

Weigel, P. & Larher, F. (1985). Occurrence of high-concentrations of glycine betaine in the glycophytic Chenopodiaceae. *Comptes Rendus de l' Academie des Sciences Serie III – Sciences de la Vie*, **300**, 65.

Weigel, P., Lerma, C. & Hanson, A.D. (1988). Choline oxidation by intact spinach chloroplasts. *Plant Physiology*, **86**, 54–60.

Weigel, P., Weretilnyk, E.A. & Hanson, A.D. (1986). Betaine aldehyde oxidation by spinach-chloroplasts. *Plant Physiology*, **82**, 753–9.

Weretilnyk, E.A., Bednarek, S., McCue, K.F., Rhodes, D. & Hanson, A.D. (1989). Comparative biochemical and immunological studies of the glycine betaine synthesis pathway in diverse families of dicotyledons. *Planta*, **178**, 342–52.

Weretilnyk, E.A. & Hanson, A.D. (1988). Betaine aldehyde dehydro-genase polymorphism in spinach – genetic and biochemical-characterization. *Biochemical Genetics*, **26**, 143–51.

Weretilnyk, E.A. & Hanson, A.D. (1989). Betaine aldehyde dehydro-genase from spinach leaves – purification, invitro translation of the messenger-RNA, and regulation by salinity. *Archives of Biochemistry and Biophysics*, **271**, 56–63.

Weretilnyk, E.A. & Hanson, A.D. (1990). Molecular-cloning of a plant betaine-aldehyde dehydrogenase, an enzyme implicated in adaptation to salinity and drought. *Proceedings of the National Academy of Sciences, USA*, **87**, 2745–9.

Weretilnyk, E.A. & Summers, P.S. (1992). Betaine and choline metab-olism in higher plants. In *Biosynthesis and Molecular Regulation of Amino Acids in Plants*. Ed. H. Flores, J. Shannon & B. Singh. pp. 30–9. Rockville: American Society of Plant Physiologists.

Williams, W.P., Brain, A.P.R. & Dominy, P.J. (1992). Induction of non-bilayer lipid phase separations in chloroplast thylakoid mem-branes by compatible co-solutes and its relation to the thermal stability of photosystem II. *Biochimica et Biophysica Acta*, **1099**, 137–44.

Winzor, C.L., Winzor, D.J., Paleg, L.G., Jones, G.P. & Naidu, B.P. (1992). Rationalization of the effects of compatible solutes on pro-tein stability in terms of thermodynamic nonideality. *Archives of Biochemistry and Biophysics*, **296**, 102–7.

Wolff, S.D., Yancey, P.H., Stanton, T.S. & Balaban, R.S. (1989). A simple HPLC method for quantitating major organic solutes of renal medulla. *American Journal of Physiology*, **256**, 954–6.

Wood, A.J. & Goldsbrough, P.B. (1993). Cloning and expression of a cDNA-encoding betaine aldehyde dehydrogenase (BADH) in sorghum. *Plant Physiology*, **102**, 9.

Wood, K.V., Stringham, K.J., Smith, D.L., Volenec, J.J., Hender-shot, K.L., Jackson, K.A., Rich, P.J., Yang, W.J. & Rhodes, D. (1991). Betaines of alfalfa – characterization by fast-atom-bombardment and desorption chemical ionization mass-spectrometry. *Plant Physiology*, **96**, 892–7.

Wyn Jones, R.G. (1984). Phytochemical aspects of osmotic adaptation. *Recent Advances in Phytochemistry*, **18**, 55–78.

Wyn Jones, R.G. & Storey, R. (1981). Betaines. In *The Physiology and Biochemistry of Drought Resistance in Plants*, ed. L.G. Paleg & D. Aspinall, pp. 171–204. Sydney: Academic.

Wyn Jones, R.G., Storey, R., Leigh, R.A., Ahmad, N. & Pollard, A. (1977). A hypothesis on cytoplasmic osmoregulation. In *Regu-lation of Cell Membrane Activities in Higher Plants*, eds. E. Marre & O. Ciferri, pp. 121–36. Amsterdam: Elsevier/North Holland.

Yancey, P.H., Clark, M.E., Hand, S.C., Bowlus, R.D. & Somero, G.N. (1982). Living with water stress: evolution of osmolyte systems. *Science*, **217**, 1214–22.

Yuan, Z.X., Patel, A.V., Blunden, G. & Turner, C.H. (1992). Trans-4-hydroxypipecolic acid betaine from *Lamium maculatum*. *Phytochemistry*, **31**, 4351–2.

Zhao, Y., Aspinall, D. & Paleg, L.G. (1992). Protection of membrane integrity in *Medicago sativa* L. by glycinebetaine against the effects of freezing. *Journal of Plant Physiology*, **140**, 541–3.

Zuniga, G.E., Argandona, V.H. & Corcuera, L.J. (1989). Distribution of glycine-betaine and proline in water stressed and unstressed barley leaves. *Phytochemistry*, **28**, 419–20.

Zuniga, G.E. & Corcuera, L.J. (1987a). Glycine-betaine in wilted barley reduces the effects of gramine on aphids. *Phytochemistry*, **26**, 3197–200.

Zuniga, G.E. & Corcuera, L.J. (1987b). Glycine betaine accumulation influences susceptibility of water-stressed barley to the aphid *Schizaphis graminum*. *Phytochemistry*, **26**, 367–9.

A.F. TIBURCIO, R.T. BESFORD,
A. BORRELL and M. MARCÉ

Metabolism and function of polyamines during osmotically induced senescence in oat leaves and protoplasts

Introduction

Polyamines (PAs) are biologically ubiquitous aliphatic nitrogen-containing compounds of low molecular weight and polycationic nature. The diamine putrescine (Put) and the triamine spermidine (Spd) are probably synthesized by all organisms, while eukaryotes contain the tetraamine spermine (Spm) as well (Cohen, 1971).

In plants, PAs are metabolically related to the basic amino acids arginine and ornithine and therefore also to glutamic acid, a key intermediate in nitrogen metabolism (Tiburcio, Kaur-Sawhney & Galston, 1990). The biosynthetic and degradative pathways for the formation of Put, Spd and Spm in plants are now well established; the key enzymes have been characterized, but their regulation at molecular level is still obscure (Tiburcio *et al.*, 1990; Galston & Tiburcio, 1991).

Suggested roles for PA function, and the evidence for these functions, has recently been reviewed (Tiburcio *et al.*, 1993*a*). These include membrane stabilization, free radical scavenging, effects on DNA, RNA and protein synthesis, effects on the activities of RNase, protease and other enzymes, the interaction with ethylene biosynthesis, and effects on second messengers. It was concluded that, in addition to interacting with plant hormones, PAs are able to modulate plant development through (a) fundamental mechanism(s) common to all living organisms (Tiburcio *et al.*, 1993*a*).

This chapter deals with the study of PA metabolism and function during osmotically induced senescence in *Avena sativa* L. (oat) leaves and protoplasts.

Experimental model and background

The oat leaf system constitutes an excellent experimental model system for studying biochemical and molecular aspects related to leaf senescence (Thimann, 1987). When oat leaf segments are incubated in darkness, they undergo senescence as judged by their manifestation of several metabolic changes, including an immediate rise in ribonuclease activity, followed by an increase in protease activity after 6 hours, and then by a gradual decline of chlorophyll content after 24 hours (Kaur-Sawhney & Galston, 1979). Exogenous application of PAs are able to inhibit or retard each of these events. The tetraamine Spm is more active than the triamine Spd, which is in turn more active than the diamines Put and cadaverine. Application of Spm is also more effective in preventing those senescence-related events than similar treatments with other known senescence retardants such as kinetin and cycloheximide (Kaur-Sawhney & Galston, 1979; for review see Kaur-Sawhney & Galston, 1991).

While attempting to improve the viability of oat protoplasts, it was observed that the cells aged rapidly and this was a major obstacle to cell division and proliferation (Brenneman & Galston, 1975). Leaf pretreatment or addition to the culture medium of exogenous PAs were able (a) to stabilize oat protoplasts against spontaneous or induced lysis, (b) to increase the incorporation of amino acids and nucleosides into proteins and nucleic acids in leaves and protoplasts, (c) to retard chlorophyll breakdown in leaves and protoplasts, and (d) to decrease or prevent the postexcision- and senescence-induced rise of ribonuclease and other hydrolytic enzymes (Altman, Kaur-Sawhney & Galston, 1977; Galston, Altman & Kaur-Sawhney, 1978). Of those compounds, the most active were Spd and Spm, which can also induce DNA-synthesis and limited mitosis in oat mesophyll protoplasts (Kaur-Sawhney, Flores and Galston, 1980; Kaur-Sawhney & Galston, 1991).

Although these results clearly indicate that PAs are potent anti-senescence agents, caution is needed since exogenous PAs may exhibit non-specific effects that are perhaps unrelated to the physiological role of their endogenous counterparts (Birecka *et al.*, 1984). For example, exogenous PAs, like Ca^{2+}, can associate with membrane lipids to induce membrane rigidity which could be an alternative mechanism to previous interpretations concerning the physiological effects of PAs on plant senescence (Roberts, Dumbroff & Thompson, 1986). Thus, in order to establish the precise physiological role of PAs, experiments with exogenous PA application should be reinforced by parallel *in situ* experiments in which endogenous PA levels are altered by using, for example, PA metabolic inhibitors.

To alter endogenous PA levels, specific inhibitors of polyamine oxidase (PAO) activity, and of arginine decarboxylase (ADC) activity (see Fig. 1), have been mainly used. As discussed more fully in the following sections, both treatments lead to an increase of endogenous Spd and Spm levels, exhibiting antisenescence effects similar to the

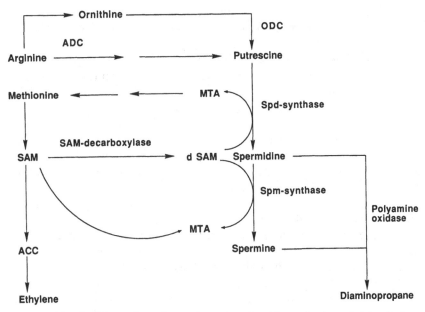

Fig. 1. Schematic pathway for polyamine biosynthesis and degradation in monocot plants.

exogenous supply of these PAs. This reinforces the view that exogenous PAs, like their endogenous counterparts, may have a specific physiological effects on oat leaf and protoplast senescence.

Research objectives

Using the experimental model system described above, two main basic objectives are currently pursued: (a) To investigate the metabolism of endogenous PAs in osmotically stressed oat leaves and (b) To understand the mechanisms of PA action on membranes and on gene expression. With regard to the practical applications of this research, various treatments involving manipulation of PA metabolism have been assessed to investigate possible antisenescence properties of PAs with oat leaves and protoplasts.

Metabolism of polyamines

Analytical methods

An improved HPLC method for the quantitative determination of dansyl-polyamines has recently been developed in our laboratory (Marcé *et al.*, in preparation).

The main advantages of the method are:

(a)   Its rapidity, since only ten minutes are needed for analysing each sample.

(b)   The excellent separation between diaminopropane (Dap) and Put. This is very convenient when working with samples of mononocot species (such as oat) which characteristically have unusual high levels of Dap.

It is difficult to measure the activity of Spd synthase (see Fig. 1), since one of the precursors, decarboxylated S-adenosylmethionine (dSAM), is unstable and not easily available. Although its synthesis has been achieved, an extinction coefficient has not been published (Yamanoha & Cohen, 1985). In the past, three general types of assay methods have been devised to assess Spd synthase activity from plant extracts. The first uses labelled Put and dSAM as substrates and depends upon the isolation of purified labelled Spd (Sindhu & Cohen, 1983). The second involves the use of methyl-labelled dSAM (which is difficult to prepare) and depends on the recovery of the labelled product, 5'-deoxy-5'-methylthioadenosine (MTA) (Sindhu & Cohen, 1983). Thirdly, the formation of Spd can be determined by a coupled enzymatic assay which avoids the use of radiolabelled substrates (Sindhu & Cohen, 1983). As discussed above, the two first methods are not reliable owing to the difficulties for preparing labelled dSAM or availability of unlabelled dSAM. Although the third method permits bypassing the difficulties with dSAM, it is tedious since it involves the purification of the enzyme polyamine oxidase (PAO) from oat leaves (Sindhu & Cohen, 1983).

To avoid most of these inconveniences, a new assay for the evaluation of Spd synthase activity has been developed (Tiburcio, Kaur-Sawhney & Galston, 1993*b*). It involves a coupled reaction and avoids the use of dSAM. The assay mixture contains SAM, [^{14}C] Put, and pyridoxal phosphate. Incorporation of [^{14}C]-label into Spd can be detected after 45 minutes of incubation at 37 °C. To assess that, first it is necessary to remove the [^{14}C]-label not incorporated into Spd by applying the samples to Dowex 50W-H$^+$ columns and eluting them with 2.3 M HCl. This treatment allows the complete elution and separation of labelled Put. Labelled Spd can be eluted from the column by using 6 M HCl. However, 6 M HCl elutes not only Spd, but also a significant fraction of [^{14}C]-labelled Spm. To avoid this problem, the use of 5 M HCl has been found to be successful, since no significant amounts of label from Spm are detected with this treatment and most of the label from Spd is also recovered (Tiburcio *et al.*, 1993*b*).

Results

By using the methods described above, as well as other well-established ones, the effect of osmotic stress on the activity of several biosynthetic enzymes and PA levels has been investigated (Tiburcio *et al.*, 1993*b*).

The results in Table 1 show that osmotically induced Put accumulation in oat leaves is due not only to an activation of the ADC pathway, but also to an inhibition of the activity of Spd synthase, the enzyme which catalyses the transformation of Put to Spd (see Fig. 1).

Table 1. *Effect of osmotic stress (6h-sorbitol) on the activity of several polyamine-biosynthetic enzymes*

Treatment	ADC	ODC	SAMDC	Spd-synthase	Put/Spd+Spm
	% in relation to control				ratio
Control	100±4	100±4	100±3	100±4.1	0.2
Sorbitol	188±19[a]	92±10	94±4	76±5[a]	1.2

[a]Significantly different from control at $P<0.05$.

Antisenescence properties of guazatine

Another possibility that could also contribute to the higher Put: (Spd + Spm) ratio in osmotically stressed oat is based on the fact that the decline of the levels of Spd and Spm is always correlated with an increase in the levels of Dap (see Fig. 1), the direct oxidation product of Spd and Spm via PAO activity (Smith, 1985).

The di-guanidine guazatine, a fungicide used as a cereal dressing which prevents germination of a wide range of pathogenic fungi, has also been shown to be a powerful inhibitor of PAO activity in oat seedlings (Smith, 1985). Although the mechanism of inhibition remains unclear, it was suggested that the active site of the enzyme is involved since a competitive relationship with the substrates Spd and Spm was observed (Smith, 1985).

The catabolic oxidation of Spd and Spm by PAO produces Dap as a common product and it appears that this is the only metabolic source of this diamine (Smith, Croker & Loeffler, 1986). Since high levels of Dap are negatively correlated with cereal leaf integrity under osmotic stress, the PAO inhibitor guazatine was used in order to prevent the oxidation of Spd and Spm (Capell, Campos & Tiburcio, 1993).

Table 2 shows that increasing concentrations of guazatine are able

Table 2. *Effect of different concentrations of guazatine on Dap/Spd+Spm ratio and total chlorophyll (Chl) levels*

Guazatine concentration (mM)	Dap/Spd+Spm ratio at 6 h	Total Chl at 48 h % in relation to control
0	2.2	100
1	0.3	270
2	0.3	300
3	0.4	230

to decrease the levels of Dap, relative to concomitant increases in the levels of the higher PAs (Spd+Spm). The lowest Dap:(Spd+Spm) ratios were obtained with 1 mM and 2 mM guazatine. These results demonstrate that it is possible to increase PA concentrations *in vivo* by supplying guazatine to oat leaves incubated in the dark and in the presence of 0.6 M sorbitol. Interestingly, incubation of oat shoots with guazatine in the light and in the absence of osmoticum does not reduce this ratio (Smith, 1985). The presence of high levels of sorbitol or mannitol in the incubation media may be important, since conversion of Spd to Dap was also reduced in oat protoplasts by using the PAO inhibitor hydroxyethylhydrazine (Mizrahi, Applewhite & Galston, 1989).

The influence of different concentrations of guazatine on Chl content of osmotically-stressed oat leaves is also shown in Table 2. The most efficient antisenescence treatments are the applications of guazatine at 2 mM and 1 mM concentrations. Furthermore, 1 mM guazatine treatment was found to be more efficient in preventing chlorophyll losses than exogenous applications of 1 mM Spd or 1 mM Spm (Capell *et al.*, 1993). Thus, the antisenescence properties of guazatine described in this study (Capell *et al.*, 1993) may be related to its ability to inhibit PAO activity. It was suggested that such inhibition prevents the immediate oxidation of endogenous Spd and Spm, consequently allowing them to exert their physiological effects (Capell *et al.*, 1993). Experiments are in progress to investigate the effect of guazatine treatments on PAO activity in osmotically stressed oat leaves.

Mechanisms of action

A suggested mechanism of plant hormone action involves attachment to a specific binding site (usually a receptor protein), and initiation of a cascade of reactions which generally lead to changes at both membrane and gene activity levels. Keeping in mind this mechanism of hormone action, PA-mediated effects on membranes and on gene expression have been investigated using the above experimental system.

Membrane level: stabilization of molecular composition and structure of thylakoid membranes

One of the mechanisms involved in the antisenescence properties of PAs may be related to their ability to maintain the integrity of cellular membranes. Thus, evidence related to the ability of PAs to stabilize the molecular composition of thylakoid membranes has been recently obtained (Besford *et al.*, 1993). This study shows the effects of exogenous PA supply and pretreatment with DL-α-difluoromethylarginine (DFMA), an irreversible inhibitor of ADC activity, on the stabilization of key polypeptides in the thylakoid membranes of osmotically stressed oat leaves using several recently produced antibody probes. Those used were site-directed antibodies to the reaction centre polypeptides of photosystem II (Dl and D2) (Besford *et al.*, 1990) and monoclonal antibodies to cytochrome f in the cytochrome b_6f complex (Besford & Thomas, 1988). Polyclonal antibodies to the large subunit (LSU) of Rubisco (Besford, 1990) were also used to assess degradation and loss of this protein from oat leaf tissue.

The results have shown that high protein levels, as analysed by SDS-PAGE, were maintained in thylakoid membranes of leaf tissue incubated in the dark in the presence of 0.6 M sorbitol when pre-treated with DFMA (which enhances endogenous Spd and Spm levels after osmotic treatment; see the following sections for more details). At 48 h incubation, the level of the thylakoid protein Dl, at the core of photosystem II, was higher in the DFMA pre-treated leaves as was the stromal protein Rubisco (as indicated by the level of large subunit, LSU). Applications of Spd, Spm or Dap were effective in retarding the loss of D1, D2 and cytochrome f from the thylakoid membranes as well as LSU and Chl from the leaf tissue.

Based on this immunological evidence, PAs were shown to stabilize the composition of the thylakoid membrane, adding further support that PAs, especially Spm, Spd and Dap, play a key role in preserving the integrity of these membranes (Besford *et al.*, 1993). Ultrastructural observations made with the electron microscope have supplied additional support to the above study. Thus, leaf segments incubated for 90 hours in the presence of Spm showed higher thylakoid membrane integrity when compared with control tissue. The application of immunocytochemical techniques using Lowicryl ultrathin sections (Testillano *et al.*, 1991) has also permitted the *in situ* localization of the D1 protein in the thylakoids of the PA-stabilized oat material.

Membrane level: inhibition of lipid peroxidation

The mechanism(s) of membrane stabilization by PAs probably involves direct binding of the protonated amino groups of PAs to the acidic phospholipids and negatively charged protein residues of membranes. As a result, PAs may affect several properties of membranes, including stabilization, activity of membrane-bound enzymes, transport and assembly of membranes (Schuber, 1989).

In plants, direct interrelationships between binding of PAs to specific membrane components and the resulting stabilization effect have not been studied. However, it has already been proposed that other indirect mechanisms may also contribute to the PA stabilizing effects on membranes. Such mechanisms may include prevention of lipid peroxidation (Dumbroff, 1991) and proteolytic attack (Kaur-Sawhney *et al.*, 1982), and inhibition of ethylene synthesis through inhibition of ACC synthase activity (Fuhrer *et al.*, 1982; Drolet *et al.*, 1986; Winer and Apelbaum, 1986) by restricting accumulation of ACC-synthase transcripts (Li *et al.*, 1992).

Lipoxygenase is likely to be involved in membrane lipid peroxidation during plant senescence (Lynch, Sridhara & Thompson, 1985). This enzyme incorporates molecular oxygen into linoleic and linolenic acids forming different lipid peroxides (Axelrod, Cheesbrough & Laakso, 1981). These products further metabolize to produce malondialdehyde, volatile hydrocarbons such as ethane and pentane, jasmonic acid, oxygen free radicals and so on (Leshem, 1988). All these compounds have been shown to be associated with plant senescence processes (Leshem, 1987).

To investigate whether the interaction of PAs with membranes would involve prevention of lipid peroxidation, the levels of malondialdehyde in osmotically stressed oat leaf tissue have been determined (Carbonell *et al.*, unpublished observations). Exogenous Spd and Spm as well as guazatine treatment significantly inhibit the formation of malondialdehyde and delay senescence. The results obtained by Western blot analysis using specific antibody probes to lipoxygenase have given further support to the contention that inhibition of lipid peroxidation by PAs may be one of the mechanisms responsible for their antisenescence effects in dark-incubated oat leaves (Borrell *et al.*, unpublished observations).

The results of this investigation provide *in vivo* support to earlier observations obtained with cell-free systems. For instance, in rat liver microsomes it was suggested that inhibition by Spm of lipid peroxidation was due to direct binding of Spm to microsomal phospholipids (Kitada *et al.*, 1979). More recently, a possible molecular mechanism by which Spm may inhibit lipid peroxidation has been advanced using vesicles containing phospholipids (Tadolini, 1988). It is related to the ability of Spm to form a ternary complex with iron and the phospholipid polar head may change the susceptibility of Fe^{2+} to auto-oxidation and thus its ability to generate free oxygen radicals (Tadolini, 1988).

The potential for PAs to act as free radical scavengers has also been recognized (Drolet *et al.*, 1986). Thus, PAs could effectively scavenge O^{2-} generated in both chemical and enzymatic cell-free systems. At concentrations of 5 to 50 mM, PAs also inhibited the production of O^{2-} by senescing microsomal membranes and the superoxide-dependent conversion of 1-aminocyclopropane-1-carboxylic acid to ethylene (Drolet *et al.*, 1986). However, later studies have questioned the postulated radical-scavenging properties of PAs (Bors *et al.*, 1989).

The experiments performed with the oat-leaf system suggest that the use of *in vivo* approaches may help to elucidate these contradictory data. Studies are in progress to characterize further the mode of PA action on lipid peroxidation of senescent oat leaves and oat mesophyll protoplasts.

Gene activity level: control by polyamines of ADC gene expression in osmotically stressed oat leaves

Activation of genes represents a large amplification process because repeated transcription of DNA into mRNA, followed by translation of the mRNA into enzymes with high catalytic activity can lead to many copies of an important cellular product. There are various control points in the flow of genetic information from DNA to a molecular product. One of these, perhaps the most important, occurs at the level of transcription. Other control points concern processing of the mRNA, then either translation on ribosomes or degradation by ribonucleases. After translation into protein, post-translational modification of the protein can occur by processes such as phosphorylation, methylation, glycosidation, and so on. There is now no doubt that all these processes controlling gene activity can be affected by plant hormones (for review see Salisbury & Ross, 1991).

Experiments have been performed attempting to address the question of whether or not changes in ADC activity induced by osmotic treatment are accompanied by parallel changes of ADC gene activity at different levels of regulation: transcription, translation and post-translation. Furthermore, in connection with the antisenescence properties of exogenous PAs in osmotically stressed oat leaves, experiments have been performed to investigate if PAs, in addition to acting at the membrane level, are also able to influence ADC gene expression.

Previous work on the same system showed that excised oat leaf pieces responded to osmotic stress within a few minutes by a marked rise in ADC activity and Put content (Flores, Young & Galston, 1985). Through the use of cycloheximide, it was shown that protein synthesis was necessary for this response to occur, and through the use of DFMA it was possible to establish that ADC activity is responsible for the rise in Put level (Flores *et al.*, 1985).

The availability of a cDNA clone encoding oat ADC (Bell & Malmberg, 1990), has allowed the investigation of changes in the levels of ADC mRNA in osmotically stressed oat leaves. Under senescence-inducing conditions, that is in the absence of exogenous PAs, the ADC mRNA levels show a significant increase after 1 hour of incubation, but then the mRNA levels rapidly decrease after 2 hours of incubation in the dark. In contrast, under antisenescence-inducing conditions, that is in the presence of exogenous Spm, there is a more marked increase of the ADC mRNA levels after 1 hour of incubation but the levels remain high after 2 hours. In the absence of PAs, the temporal increase of the ADC mRNA levels is positively correlated with a significant enhancement of the ADC enzyme activity. However, in sharp contrast, in the presence of Spm the increase of the levels of ADC mRNA is not accompanied by a parallel increase of the ADC enzyme activity. In fact, the levels of ADC activity are even lower than the controls (Tiburcio *et al.*, unpublished observations). To solve this problem, experiments analysing changes at the protein level were needed to understand the mechanism of regulation of ADC gene expression in this particular system.

Three different types of antibodies specific for ADC were developed by Malmberg *et al.* (1992). Murine monoclonal antibody 'As-8B5' which immunoprecipitates oat ADC enzyme activity, but does not work well on Western blots. Chicken polyclonal antibody 'anti-C' raised against the 22% of the carboxyl end of oat ADC that reacts with a 24 kD polypeptide and also with a 34 kD polypeptide in the same Western blot. Thirdly, a chicken polyclonal antibody 'anti-N' raised against the amino 33% of oat ADC that reacts with a polypeptide of 42 kD and with another polypeptide of about 60 kD (Malmberg *et al.*, 1992).

By using these immunological probes, Malmberg *et al.* (1992) proposed that ADC in oat is first synthesized in the form of an inactive precursor protein of 66 kD which is then cleaved to produce a 42 kD polypeptide containing the original amino terminus and a 24 kD polypeptide containing the original C-terminus. Since DFMA binds to the 24 kD fragment, it is suggested that the active site of the ADC enzyme must reside in this fragment which consequently is considered as the active ADC form (Malmberg *et al.*, 1992). According to this model, the development of antibodies recognizing both the pre-protein and the active ADC form in the same Western blot would be very useful to understand the mechanism of enzyme regulation.

Having this in mind, site-directed polyclonal antibodies specific to ADC have recently been developed (unpublished). The strategy was as follows: an amino acid sequence near to the C-terminus deduced from the nucleotide sequence of the ADC gene (Bell & Malmberg, 1990) was selected. This polypeptide of 20 amino acids has been synthesized, coupled to PPD (purified-protein-derivative of tuberculin) carrier protein and injected into rabbits to produce the corresponding antibodies. The technique is described more fully in Besford *et al.* (1990). The choice of this polypeptide was based on the fact that this region does not contain substrate binding sites and was predicted to be antigenic.

Proteins from the different oat treated samples have been extracted, separated by SDS-PAGE and analysed by Western blotting. The results reveal a significant increase in the levels of the 24 kD polypeptide (ADC active form) in osmotically stressed oat leaves when compared with similar samples incubated in the absence of osmoticum for 4 and 24 hours. Only the 24 kD band is detected when the experiments are performed in the absence of exogenous Spm.

In contrast, the presence of exogenous Spm leads to a dramatic decrease in the levels of the 24 kD band from 4 to 24 hours of incubation. No 24 kD band is detected at 24 and 48 hours, instead an accumulation of a polypeptide of about 60 kD is observed. This band (corresponding to the inactive ADC preprotein) is only detected in the presence of Spm (or Spd) after 24 and 48 hours of incubation.

These results indicate that the recently developed site-directed antibodies are able to detect both the ADC precursor protein and the active form. Furthermore, the results also explain the decrease of ADC activity in the leaves incubated in the presence of Spm, in spite of the increased levels of ADC mRNA. The evidence supports the view that exogenous Spm inhibits the post-translational processing (clipping) of the precursor protein which results in increased levels of this inactive ADC form with a consequent decrease of the active ADC form. This represents a new model of post-translational regulation by Spm of ADC gene expression.

Although the mechanism of post-translational control of nascent ADC protein by Spm remains unclear, it is feasible that a protease may be involved in the reaction. At present, it can be speculated that inhibition of a specific protease by Spm could be a possible mechanism involved in this post-translational control. Further work is needed to confirm this hypothesis.

Practical applications: improvement of oat mesophyll protoplasts

Genetic factors and injuries (excision and wounding) sustained during isolation of mesophyll protoplasts have been emphasized in the past as the main reasons for lack of sustained cell division and failure of plant regeneration in cereals (Flores, Kaur-Sawhney & Galston, 1981). For example, it has been suggested (Fuchs & Galston, 1976) that morphological instability and deterioration of oat mesophyll protoplasts may be related to a complex of senescence-induced changes, especially RNase activity (Kaur-Sawhney *et al.*, 1976), following leaf excision and protoplast isolation. However, tobacco protoplasts which readily divide also show a dramatic increase in RNase activity after 24 h of isolation (Lazar *et al.*, 1973), and this increase has been shown to be due mostly to osmotic stress rather than to injury caused during isolation of protoplasts (Premecz *et al.*, 1977).

We have formulated the hypothesis that difficulties of monocot mesophyll protoplasts to retain viability compared with those derived from dicot protoplasts may be due to their different PA metabolism in response to physiological stress incurred during isolation in high osmotica (Tiburcio *et al.*, 1986*a,b*).

As discussed in the preceding sections, dark-incubation of oat leaves in the presence of 0.6 M sorbitol results in a dramatic increase in the levels of Put, although the levels of Spd and Spm decrease. This response is the opposite to that observed in leaves of dicot genera which readily regenerate plants from mesophyll protoplasts (Tiburcio *et al.*, 1986*a*). In the solanaceous species (*Nicotiana* and *Capsicum*) the osmotically induced increase in higher PAs is due mainly to Spd, whereas the legumes *Vigna* and *Trigonella* increase their Spm levels after osmotic treatment. A similar pattern was observed when comparing freshly isolated oat protoplasts with readily regenerating dicot protoplasts (Tiburcio *et al.*, 1986*a*).

The high Put:(Spd+Spm) ratio in leaves and protoplasts of species which yield poorly dividing protoplasts contrasts with the low ratio in leaves and protoplasts of species which yield readily dividing protoplasts. This is interesting since many investigations point to an important role of the higher PAs (Spd and Spm) in DNA synthesis and mitosis (Heby, 1981). Conversely, high concentrations of exogenously applied or endogenous Put appear toxic to plant cells (Guarino & Cohen, 1979; Strogonov, Shevyakova & Kabanov, 1972). Therefore, we suggested that this differing response of PA metabolism to osmotic stress may account in part for the failure of cereal mesophyll protoplasts to develop readily *in vitro* (Tiburcio *et al.*, 1986*a*).

To test this hypothesis, we attempted to prevent the accumulation of Put in oat protoplasts by pre-treating oat leaves with DFMA (Tiburcio *et al.*, 1986*b*). Leaf pretreatment with 10 mM DFMA before a 6 h osmotic treatment caused a small decrease of Put and a 2-fold increase of Spm levels. After 136 h of osmotic stress, Put levels in DFMA-pretreated leaves increased a small amount, but Spd and Spm levels increased dramatically (Tiburcio *et al.*, 1986*b*). Furthermore, the osmotically induced inhibition of Spd-synthase activity was prevented by the pretreatment of the leaves with DFMA (Tiburcio *et al.*, 1993*b*). The increase in the titres of Spd and Spm may be associated with the reduced chlorophyll loss and the enhanced ability of DFMA-pretreated leaves to incorporate tritiated thymidine, uridine and leucine into macromolecules (Tiburcio *et al.*, 1986*b*). The higher net macromolecular synthesis and lowered rates of senescence observed in DFMA-pretreated leaves after osmotic treatment encouraged us to examine whether protoplasts isolated from such leaves will grow better in culture. As we expected, protoplasts isolated from DFMA-pretreated leaves appeared greener, with greater cytoplasmic streaming, more uniform regeneration of cell walls, and showing less lysis than protoplasts extracted from control leaves. The overall viability and cell division of the protoplasts was significantly improved by the DFMA pretreatment, thus supporting the hypothesis that the osmotically induced rise in Put and blockage of its conversion to Spd and Spm may contribute to the lack of sustained cell division in cereal mesophyll protoplasts (Tiburcio *et al.*, 1986*b*).

The DFMA pretreatment, which increases the endogenous levels of Spd and Spm, was clearly superior in improving oat mesophyll protoplast cultures than previous reported treatments using exogenous applications of PAs (Kaur-Sawhney *et al.*, 1980). In our hands, the mesophyll oat protoplasts pretreated with DFMA were stable, not undergoing the phenomenon of lysis often encountered with cereal mesophyll protoplasts, and many of them were able to regenerate complete cell walls, as opposed to the usual patchy deposition of cellulose (Kinnersley, Racusen & Galston, 1978). One-month DFMA-pretreated cultures showed an increase in all stages of protoplast development (see Fig. 2), when compared with control protoplasts (Tiburcio *et al.*, 1986*b*). While control cultures had all senesced, most of the DFMA-pretreated cultures were maintained for up to 2 months. Such cultures when transferred to solid media were also capable of microcolony formation (Tiburcio *et al.*, 1986*b*).

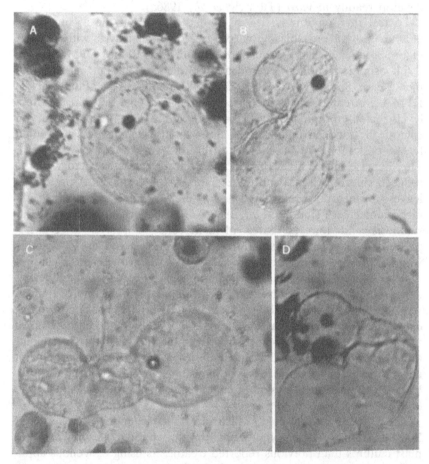

Fig. 2. Development of oat mesophyll protoplasts pretreated with DFMA. (A) 'Oval' stage; (B) 'Double'; (C) 'Triple'; (D) 'Multiple', clusters of cells.

In current experiments we are assaying different treatments and conditions in order to further improve the cell division of oat mesophyll protoplasts. One of the approaches has been the use of the inhibitor guazatine. The antisenescence properties of guazatine in osmotically stressed oat leaves have been described in a preceding section. Experiments are now in progress to investigate whether or not guazatine treatments are also able to improve the viability and cell division of oat mesophyll protoplasts.

Future perspectives

In this chapter, it has been shown that PAs act as antisenescence compounds with mechanisms of action at the membrane and gene activity levels similar to those of the plant hormones. It is known that plant hormones might utilize the phosphatidylinositol (PI) cycle for the transduction of signals between plant cells. This contention is supported by effects on PI metabolites and/or PI turnover in response to auxins, cytokinins and gibberellic acid (Morre *et al.*, 1984; Ettlinger & Lehle, 1988; Boss & Morre, 1989).

The process begins with the binding of the hormone to a receptor protein in the plasma membrane of a target cell. The bound hormone–receptor complex activates a nearby membrane enzyme called phospholipase c (PLC). This enzyme PLC then hydrolyses a membrane lipid, phosphatidylinositol-4,5-bisphosphate (PIP2) to release two secondary messengers: inositol-1,4,5-triphosphate (IP3) and diacylglycerol (DAG). These compounds have further activity and can cause a cascade of responses. Thus, IP3 moves to the tonoplast, where it combines with a receptor that activates a Ca^{2+} pump or transporter that moves Ca^{2+} from the vacuole to the cytosol. DAG, which remains membrane bound, activates protein kinase c (PKC). PKC is also activated by Ca^{2+} released from the vacuole, so various enzymes become phosphorylated by PKC. Calcium also activates other protein kinases and other enzymes, when free or bound with calmodulin. Therefore, a primary hormonal stimulus finally leads to modified enzyme activity and altered metabolic and physiological processes. To complete the PI cycle, IP3 loses phosphates by hydrolysis to form IP2 and IP, which is then converted back to phosphatidylinositol (PI) and other phosphoinositide lipids (PIP and PIP2) in the plasma membrane (for review see Salisbury & Ross, 1991).

To know which components of the signal transduction pathway are affected by PAs, experiments involving interaction of PAs with several hormones and secondary messengers should be performed. In this sense, the effect of hormones and Spd on the turnover of inositolphospholipids in *Brassica* hypocotyls has been recently investigated (Dureja-Munjal, Acharya & Guha-Mukherjee, 1992). When hypocotyls were incubated in auxin and cytokinin solutions, significant decrease in the levels of PIP2, PIP and PI were observed. On the other hand, IP2 and IP3 levels increased over the control. In contrast, when hypocotyls were incubated in Spd solutions, the levels of PIP and PIP2 increased compared with both control or hormone-treated hypocotyls. On the other hand, Spd caused a decline of the levels of PI, IP2 and IP3. Auxin and cytokinin act mainly on PIP2 and induce its breakdown to produce IP2 and IP3; whereas Spd may stimulate PIP kinases thus maintaining high levels of PIP and PIP2 (Dureja-Munjal *et al.*, 1992). Although Spd and hormones may act at different levels, these results suggest that PAs, like hormones, may also utilize the PI cycle for the transduction of signals between plant cells.

The recent availability of cDNA coding for ADC (Bell & Malmberg, 1990) and other molecular probes being developed, allow the techniques of molecular biology to be used. For example, as shown in the present article, the application of molecular probes increases the knowledge of the effects of PAs on specific gene expression and post-transcriptional control. Also, the study at the molecular level of interactions between PAs, hormones and secondary messengers will allow in the future the understanding of the signal transduction mechanisms involved in the prevention by PAs of osmotically induced senescence in oat leaves and protoplasts.

Finally, by using transgenic plants over-or under-expressing genes coding for PA metabolism, it should be possible to test directly the role of PAs in different developmental and metabolic processes. This approach has recently been used with success to overproduce the diamine cadaverine in transgenic *Nicotiana tabacum* plants (Herminghaus *et al.*, 1991) or Put-derived alkaloids in transgenic roots of *Nicotiana rustica* (Hamill *et al.*, 1990).

Acknowledgements

We thank Dr Malmberg for the gift of the ADC cDNA clone. The research has been supported by grants CICYT-BIO 90-130, CICYT-BIO 93-130, CICYT-PTR-91-70 (AFT), MEC Acciones Integradas HB-79 and HB-119A (AFT and RTB), MEC Acciones Integrada HI-112 (AFT and N. Bagni) and AFRC, UK (RTB). The contribution of our co-workers, T. Capell, X. Figueras, J. Campos, M. Risueno,

C. Richardson, P. Testillano, D. Wilkins and L. Carbonell is gratefully acknowledged.

References

Altman, A., Kaur-Sawhney, R. & Galston, A.W. (1977). Stabilization of oat protoplasts through polyamine-mediated inhibition of senescence. *Plant Physiology*, **60**, 570–4.

Axelrod, B., Cheesbrough, T.M. & Laakso, S. (1981). Lipoxygenase from soybeans. In *Methods in Enzymology*, ed. J.M. Lowenstein, pp. 441–51. New York: Academic Press.

Bell, E. & Malmberg, R.L. (1990). Analysis of a cDNA encoding arginine decarboxylase from oat reveals similarity to the *Escherichia coli* arginine decarboxylase and evidence protein processing. *Molecular and General Genetics*, **224**, 431–6.

Besford, R. (1990). The greenhouse effect: acclimation of tomato plants growing in high CO_2, relative changes in Calvin cycle enzymes. *Journal of Plant Physiology*, **136**, 458–63.

Besford, R. & Thomas, B. (1988). Production of monoclonal antibodies to thylakoid proteins. In *Meeting on Photosynthesis*, pp. 50. London: AFRC.

Besford, R., Thomas, B., Huskisson, N. & Butcher, G. (1990). Characterization of conformers of Dl of photosystem II using site-directed antibodies. *Zeitschrift für Naturforschung*, **45**, 621–6.

Besford, R., Richardson, C., Campos, J.L. & Tiburcio, A.F. (1993). Effect of polyamines on stabilization of molecular complexes in thylakoid membranes of osmotically stressed oat leaves. *Planta*, **189**, 201–6.

Birecka, H., Di Nolfo, T.E., Martin, W.B. & Frohlich, M.W. (1984). Polyamines and leaf senescence in pyrrolizidine alkaloid-bearing *Heliotropium* plants. *Phytochemistry*, **23**, 991–7.

Bors, W., Langebartels, C., Michel, C. & Sandermann, H. Jr. (1989). Polyamines as radical scavengers and protectants against ozone damage. *Phytochemistry*, **28**, 1589–95.

Boss, W.F. & Morre, D.J., eds. (1989). *Second Messengers in Plant Growth and Development*, New York: Alan R. Liss.

Brenneman, F.M. & Galston, A.W. (1975). Experiments on the cultivation of protoplasts and calli of agriculturally important plants. *Biochemie und Physiologie der Pflanzen*, **168**, 453–71.

Capell, T., Campos, J.L. & Tiburcio, A.F. (1993). Antisenescence properties of guazatine in osmotically stressed oat leaves. *Phytochemistry*, **32**, 785–8.

Cohen, S.S. (1971). *Introduction to Polyamines*. New York: Prentice-Hall, Englewood Cliffs.

Drolet, G., Dumbroff, E.B., Legge, R.L. & Thompson, J.E. (1986). Radical scavenging properties of polyamines. *Phytochemistry*, **25**, 367–71.

Dumbroff, E. (1991). Mechanisms of polyamine action during plant development. In *Lecture Course on Polyamines as Modulators of Plant Development,* ed. A.W. Galston & A.F. Tiburcio, pp. 62–6. Madrid: Fundacion Juan March.

Dureja-Munjal, I., Acharya, M. & Guha-Mukherjee, S. (1992). Effect of hormones and spermidine on the turnover of inositolphospholipids in *Brassica* seedlings. *Phytochemistry,* **31,** 1161–3.

Ettlinger, C. & Lehle, L. (1988). Auxin induces rapid changes in phosphatidylinositol metabolites. *Nature. London,* **331,** 176–8.

Flores, H.E., Kaur-Sawhney, R. & Galston, A.W. (1981). Protoplasts as vehicles for plant propagation and improvement. In *Advances in Cell Culture,* ed. K. Maramorosch, Vol 1, pp. 241–79. New York: Academic Press.

Flores, H.E., Young, N.D. & Galston, A.W. (1985). Polyamine metabolism and plant stress. In *Cellular and Molecular Biology of Plant Stress,* ed. J.L. Key & T. Kosuge, pp. 93–114. New York: Alan R. Liss.

Fuhrer, J., Kaur-Sawhney, R., Shih, L.-M. & Galston, A.W. (1982). Effect of exogenous 1,3-diaminopropane and spermidine on senescence of oat leaves. *Plant Physiology,* **70,** 1597–600.

Fuchs, Y. & Galston, A.W. (1976) Macromolecular synthesis in oat leaf protoplasts. *Plant and Cell Physiology,* **17,** 475–82.

Galston, A.W., Altman, A. & Kaur-Sawhney, R. (1978). Polyamines, ribonuclease and the improvement of oat leaf protoplasts. *Plant Science Letters,* **11,** 69–79.

Galston, A.W & Tiburcio, A.F., eds. (1991). *Lecture Course on Polyamines as Modulators of Plant Development,* Vol 257, Madrid: Fundacion Juan March.

Guarino, L.A. & Cohen, S.S. (1979). Mechanism of toxicity of putrescine in *Anacystis nidulans. Proceedings of the National Academy of Sciences, USA,* **76,** 3660–4.

Hamill, J.D., Robins, R.J., Parr, A.J., Evans, D.M., Furze, J.M. & Rhodes, M.J.C. (1990). Over-expressing a yeast ornithine decarboxylase gene in transgenic roots of *Nicotiana rustica* can lead to enhanced nicotine accumulation. *Plant Molecular Biology,* 15, 27–38.

Heby, O. (1981). Role of polyamines in the control of cell proliferation and differentiation. *Differentiation,* **19,** 1–20.

Herminghaus, S., Schreier, P.H., McCarthy, J.E.G., Landsmann, J., Bottelman, J. & Berlin, J. (1991). Expression of a bacterial lysine decarboxylase gene and transport of the protein into chloroplasts of transgenic tobacco. *Plant Molecular Biology,* **17,** 475–86.

Kaur-Sawhney, R., Rancillac, M., Staskawicz, B., Adams, W. & Galston, A.W. (1976). Effect of cycloheximide and kinetin on yield, integrity and metabolic activity of oat leaf protoplasts. *Plant Science Letters,* **7,** 57–67.

Kaur-Sawhney, R. & Galston, A.W. (1979). Interaction of polyamines pand light on biochemical processes involved in leaf senescence. *Plant Cell and Environment*, **2**, 189–96.

Kaur-Sawhney, R. & Galston, A.W. (1991). Physiological and biochemical studies on the anti-senescence properties of polyamines in plants. In *Biochemistry and Physiology of Polyamines in Plants*, ed. R.D. Slocum & H.E. Flores, pp. 201–11. Boca Raton: CRC Press.

Kaur-Sawhney, R., Flores, H.E. & Galston, A.W. (1980). Polyamine-induced DNA synthesis and mitosis in oat leaf protoplasts. *Plant Physiology*, **65**, 368–71.

Kaur-Sawhney, R., Shih, L., Cegielska, T. & Galston, A.W. (1982). Inhibition of protease activity by polyamines. Relevance for control of leaf senescence. *FEBS Letters*, **145**, 345–9 .

Kinnersley, A., Racusen, R. & Galston, A.W. (1978). A comparison of regenerated cell walls in tobacco and cereal protoplasts. *Planta*, **139**, 155–8.

Kitada, M., Igarashi, K., Hirose, S. & Kitagawa, H. (1979). Inhibition by polyamines of lipid peroxide formation in rat liver microsomes. *Biochemical and Biophysical Research Communications*, **87**, 388–94.

Lazar, G., Borbely, G., Udvardy, J., Premecz, G. & Farkas, G. (1973). Osmotic shock triggers an increase in ribonuclease level in protoplasts isolated from tobacco leaves. *Plant Science Letters*, **1**, 53–7.

Leshem, Y.Y. (1987). Membrane phospholipid catabolism and Ca^{2+} activity in control of senescence. *Physiologia Plantarum*, **69**, 551–9.

Leshem, Y.Y. (1988). Plant senescence processes and free radicals. *Free Radical Biology and Medicine*, **5**, 39–49.

Li, N., Parsons, B., Liu, D. & Mattoo, A. (1992). Accumulation of wound-inducible ACC synthase transcript in tomato fruit is inhibited by salicylic acid and polyamines. *Plant Molecular Biology*, **18**, 477–87.

Lynch, D.V., Sridhara, S. & Thompson, J.E. (1985). Lipoxygenase generated hydroperoxides account for the nonphysiological features of ethylene formation from 1-aminocyclopropane- 1-carboxylic acid by microsomal membranes of carnations. *Planta*, **164**, 121–5.

Malmberg, R.L., Smith, K.E., Bell, E. & Cellino, M.L. (1992). Arginine decarboxylase of oats is clipped from a precursor into two polypeptides found in the soluble enzyme. *Plant Physiology*, **100**, 146–52.

Mizrahi, Y., Applewhite, P.B. & Galston, A.W. (1989). Polyamine binding to proteins in oat and petunia protoplasts. *Plant Physiology*, **91**, 738–43.

Morre, D.J., Gripshover, B., Monroe, A.L. & Morre, J.T. (1984). Phosphatidylinositol turnover in isolated soybean membranes stimulated by the synthetic growth hormone 2,4dichlorophenoxyacetic acid. *Journal of Biological Chemistry*, **259**, 15364–8.

Premecz, G., Ruzicska, P., Olah, T., Guyas, A., Nyitrai, A., Palfi, G. & Farkas, G. (1977). Is the increase in ribonuclease level in isolated tobacco protoplasts due to osmotic stress? *Plant Science Letters*, **9**, 195–200.

Roberts, D.R., Dumbroff, E.B. & Thompson, J.E. (1986). Exogenous polyamines alter membrane fluidity in bean leaves – a basis for potential misinterpretation of their true physiological role. *Planta*, **167**, 395–401.

Salisbury, F.B. & Ross, C.W. (1991). *Plant Physiology*, Belmont: Wadsworth Publishing Company.

Schuber, F. (1989). Influence of polyamines on membrane functions. *Biochemical Journal*, **260**, 1–10.

Sindhu, R.K. & Cohen, S.S. (1983). Putrescine aminopropyltransferase (spermidine synthase) of Chinese cabbage. In *Methods in Enzymology*, ed. H. Tabor & C.W. Tabor, Vol. 94, pp. 279–85. New York: Academic Press.

Smith, T.A. (1985). The inhibition and activation of polyamine oxidase from oat seedlings. *Plant Growth Regulation*, **3**, 269–75.

Smith, T.A., Croker, S.J. & Loeffler, R.S.T. (1986). Occurrence in higher plants of 1-(3-aminopropyl)-pyrrolinium and pyrroline: products of polyamine oxidation. *Phytochemistry*, **25**, 683–9.

Strogonov, B.P., Shevyakova, N.I. & Kabanov, Y.Y. (1972). Diamines in plant metabolism under conditions of salinization. *Fiziologyia Rastenii (English Translation) Plant Physiology*, **19**, 1098–104.

Tadolini, B. (1988). Polyamine inhibition of lipoperoxidation. The influence of polyamines on iron oxidation in the presence of compounds mimicking phospholipid polar heads. *Biochemical Journal*, **249**, 33–6.

Testillano, P.S., Sanchez-Pina, M.A., Olmedilla, A., Ollacarizqueta, M.A. & Risueño, M.C. (1991). A specific ultrastructural method to reveal DNA: the NAMA-Ur. *Journal of Histochemistry and Cytochemistry*, **39**, 1427–38.

Thimann, K.V. (1987). Plant senescence: a proposed integration of the constituent processes. In *Plant Senescence: Its Biochemistry and Physiology*, ed. W.W. Thomson, E.A. Nothnagel & R.C. Huffaker, pp. 1–19. Rockville: American Society of Plant Physiologists.

Tiburcio, A.F., Masdeu, M.A., Dumortier, F.M. & Galston, A.W. (1986a). Polyamine metabolism and osmotic stress: I. Relation to protoplast viability. *Plant Physiology*, **82**, 369–74.

Tiburcio, A.F., Kaur-Sawhney, R. & Galston, A.W. (1986b). Polyamine metabolism and osmotic stress: II. Improvement of oat protoplasts by an inhibitor of arginine decarboxylase. *Plant Physiology*, **82**, 375–8.

Tiburcio, A.F., Kaur-Sawhney, R. & Galston, A.W. (1990). Polyamine metabolism. In *The Biochemistry of Plants, Intermediary Nitrogen Metabolism*, ed. B.J. Miflin & P.J. Lea, pp. 283–325. New York: Academic Press.

Tiburcio, A.F., Campos, J.L., Figueras, X. & Besford, R.T. (1993a). Recent advances in the understanding of polyamine functions during plant development. *Plant Growth Regulation*, **12**, 331–40.

Tiburcio, A.F., Kaur-Sawhney, R. & Galston, A.W. (1993*b*). Spermidine biosynthesis as affected by osmotic stress in oat leaves. *Plant Growth Regulation*, **13**, 103–9.

Yamanoha, B. & Cohen, S.S. (1985). S-adenosylmethionine decarboxylase and spermidine synthase from Chinese cabbage. *Plant Physiology*, **78**, 784–90.

Winer, L. & Apelbaum, A. (1986). Involvement of polyamines in the development and ripening of avocado fruits. *Journal of Plant Physiology*, **126**, 223–33.

OLE SIBBESEN, BIRGIT MARIA KOCH,
PIERRE ROUZÉ,
BIRGER LINDBERG MØLLER
and BARBARA ANN HALKIER

Biosynthesis of cyanogenic glucosides. Elucidation of the pathway and characterization of the cytochromes P-450 involved

Introduction

Cyanogenic glucosides are secondary plant products which, upon hydrolysis, release HCN. Cyanogenic glucosides are found in more than 2000 plant species, including the agriculturally important sorghum and cassava. Insufficient removal of the cyanogenic glucosides present in the cassava tubers constitutes a potential health hazard for the millions of people who are dependent on these tubers as their staple food. The presence of cyanogenic glucosides in cassava tubers have been shown to improve their resistance against the cassava root borer (Bellotti & Arias, 1993). However, in other cyanogenic plants, the cyanide released may be more harmful to the host plant than to the pest organism. This is the case in the cyanogenic rubber tree (*Hevea brasiliensis*), where elevated amounts of cyanogenic glucosides result in increased sensitivity to attack by the fungus *Microcyclus ulei* (Lieberei, 1986). Similarly, the presence of the cyanogenic glucoside *epi*-heterodendrin in the epidermal cells of barley seedlings is correlated with an increased sensitivity to the mildew fungus *Erysiphe graminis* Pourmohseni & Ibenthal, 1991. In these cases, the HCN released upon infection is impairing the plant defence responses, either by inhibiting the synthesis of phytoalexins or by inhibiting polyphenol oxidases (Lieberei *et al.* 1989). The presence of even minor amounts of *epi*-heterodendrin in barley malt causes a problem in the brewing industry (Cook *et al.* 1990). During the distilling process cyanide is liberated, which results in the formation of the carcinogenic ethylcarbamate, formed from cyanide and ethanol. Modulating the level of cyanogenic glucosides in specific tissues could potentially provide increased disease and pest resistance as well as improved nutritional value of foods. A prerequisite for specific manipulations of the level of cyanogenic glucosides is the identification and characterization of the enzymes (and genes) involved in the biosynthesis. To date, the most detailed studies on the biosynthesis of cyanogenic glucosides has been carried out in various lines of *Sorghum* (Møller & Poulton, 1993). Based on the similarity of the biosynthetic reactions in different cyanogenic plants, it is believed that the information obtained in the sorghum system can be extrapolated to other cyanogenic plants.

The biosynthetic pathway

The primary precursors of cyanogenic glucosides are restricted to the five hydrophobic protein amino acids valine, leucine, isoleucine, phenylalanine and tyrosine, and to a single non-protein amino acid, cyclopentenylglycine. In early studies on the biosynthesis, radioactively labelled amino acids were administered to intact seedlings (Akazawa, Miljanich & Conn, 1960). Only the precursor amino acid and the cyanogenic glucoside accumulated in the seedlings. No intermediates could be identified. The biosynthetic studies were greatly facilitated by the isolation of a biosynthetically active microsomal preparation from etiolated seedlings of sorghum, which could carry out the conversion of L-tyrosine to (S)-p-hydroxymandelonitrile. This includes all but the final glycosylation reaction in the biosynthetic pathway (McFarlane, Lees & Conn, 1975). *In vivo*, (S)-p-hydroxymandelonitrile is glycosylated into dhurrin by a soluble UDPG-glucosyltransferase. *In vitro*, p-hydroxymandelonitrile is unstable and spontaneously dissociates to produce p-hydroxybenzaldehyde and HCN. Similar microsomal preparations carrying out homologous reactions have subsequently been isolated from the cyanogenic plants *Triglochinum maritima* (Hösel & Nahrstedt, 1980), *Linum usitatissimum* (flax) (Cutler & Conn, 1981), *Trifolium repens* (white clover) (Collinge & Hughes, 1982), and, most recently, *Manihot esculenta* (cassava) (Koch *et al.* 1992). By using the microsomal preparation isolated from sorghum, it has been demonstrated, that the biosynthetic pathway involves N-hydroxytyrosine, 2-nitro-3-(p-hydroxyphenyl)propionic acid, 1-*aci*-nitro-2-(p-hydroxyphenyl)ethane, *(E)*- and (Z)-p-hydroxyphenylaldehyde oxime, p-hydroxyphenylacetonitrile, and p-hydroxymandelonitrile (McFarlane, Lees & Conn, 1975; Shimada & Conn, 1977; Møller & Conn, 1979; Halkier, Olsen & Møller, 1989; Halkier & Møller, 1990; Halkier, Lykkesfeldt & Møller, 1991) (Fig.1). Except for the 2-nitro-3-(p-hydroxyphenyl)propionic acid, which is a very labile compound, all the intermediates have been shown to be produced and metabolized by the microsomal preparation.

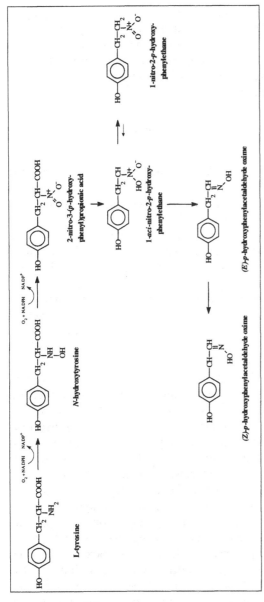

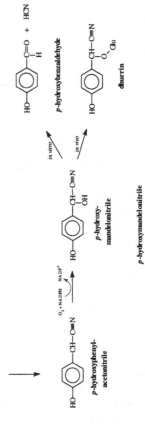

Fig. 1. The biosynthetic pathway for the cyanogenic glucoside dhurrin for sorghum. The cytochrome P-450$_{TYR}$ catalyses the reactions in the box, the conversion of tyrosine to *p*-hydroxyphenylacetaldoxime.

The biosynthetic enzyme system in sorghum constitutes a highly organized enzyme system providing an efficient mechanism for channelling of the generated intermediates between L-tyrosine and p-hydroxymandelonitrile (Møller & Conn, 1980). As the only intermediate, the (Z)-p-hydroxyphenylaldehyde oxime exchanges freely with exogenously added (Z)-p-hydroxyphenylaldehyde oxime.

Involvement of cytochrome P-450 dependent monooxygenases

Stoichiometric measurements of oxygen consumption and biosynthetic activity have indicated three hydroxylation reactions in the biosynthesis of dhurrin (Halkier & Møller, 1990). Two molecules of oxygen are consumed in two consecutive N-hydroxylation reactions in the conversion of L-tyrosine to p-hydroxyphenylacetaldehyde oxime and one oxygen molecule is consumed in a C-hydroxylation reaction converting p-hydroxyphenylacetonitrile to p-hydroxymandelonitrile (Halkier, Lykkesfeldt & Møller, 1991).

The hydroxylation reactions of L-tyrosine and p-hydroxyphenylacetonitrile are inhibited by carbon monoxide and the inhibition is reversed by 450 nm light (Halkier & Møller, 1991). This demonstrates that the L-tyrosine N-hydroxylase and the p-hydroxyphenylacetonitrile C-hydroxylase are cytochrome P-450 dependent monooxygenases. In the presence of oxygen, the metabolism of p-hydroxyphenylacetaldehyde oxime is dependent on NADPH and results in the production of p-hydroxymandelonitrile with no accumulation of p-hydroxyphenylacetonitrile in the reaction mixture. The metabolism of p-hydroxyphenylacetaldehyde oxime is photo-reversibly inhibited by carbon monoxide, by cytochrome P-450 inhibitors and by antibodies towards NADPH-cytochrome P-450 oxidoreductase (Halkier & Møller, 1991). Thus it is not possible to dissect the cofactor requirements for the conversion of p-hydroxyphenylacetaldehyde oxime to p-hydroxymandelonitrile into two separate reactions. If the results obtained for the p-hydroxyphenylacetaldehyde oxime metabolizing enzyme solely reflect the characteristics of the p-hydroxymandelonitrile monooxygenase, then the conversion of aldehyde oxime to nitrile could be a simple dehydration reaction (Halkier & Møller, 1991). However, it is surprising, that the enzyme, which dehydrates p-hydroxyphenylacetaldehyde oxime, behaves identical to the C-hydroxylase converting the nitrile to the α-hydroxynitrile at the experimental conditions tested.

Cytochromes P-450 derive their name from a characteristic absorption peak at 450 nm in the difference spectrum obtained by the addition of carbon monoxide to reduced cytochrome P-450. Cytochrome P-450 dependent monooxygenase reactions are dependent on small electron transport chains, where reducing equivalents from NADPH are transferred via a flavin-containing oxidoreductase to the terminal cytochrome P-450 monooxygenase. The cytochrome P-450 monooxygenase is the specific component, which binds the substrate and activates molecular oxygen (Guengerich, 1991). A rapidly increasing number of cytochromes P-450 are being identified in plants. Cytochrome P-450 reactions are implicated in various biosynthetic pathways leading to several products as lignin phenolics, sterols, alkaloids and phytoalexins (Donaldson & Luster, 1991). While plant cytochromes P-450 generally have been presumed to have a narrow substrate specificity and to be mainly involved in biosynthesis of primary and secondary metabolites, animals have additionally developed a wide range of cytochromes P-450, which function in the detoxification of various compounds of exogenous origin. The incorporation of hydroxyl groups facilitates their metabolism and subsequent inactivation or their excretion. Rat liver contains approximately 30 different cytochromes P-450 with a broad substrate specificity involved in detoxification and metabolism of substances of exogenous origin. Animals also posses cytochromes P-450 with a narrow substrate specificity involved in biosynthesis of steroids, vitamins and fatty acids. In recent years, however, the involvement of plant cytochromes P-450 in metabolism of xenobiotics has been established. Cytochromes P-450 are implicated in the appearance of multiple resistance to several classes of herbicides in weeds. It is an unresolved question whether the detoxifying plant cytochromes P-450 have endogenous substrates different from the herbicides, or whether plants, like animals, have developed defense mechanisms based on cytochrome P-450 detoxification of warfare compounds from other plants, insects, microorganisms or man.

Purification of the cytochrome P-450-dependent monooxygenases

The isolation of plant cytochromes P-450 has proven difficult because generally the proteins are only produced in minute amounts and only under specific developmental stages and environmental conditions. While total cytochrome P-450 constitutes 10% of the protein in rat liver microsomes, total cytochrome P-450 in microsomes isolated from etiolated sorghum seedlings constitute less than 1%. For the isolation of a gene encoding a plant cytochrome P-450 enzyme of interest, an approach alternative to protein purification has been developed. The haem-binding domain of cytochromes P-450 is very conserved through-out this enzyme class, and the domain is situated close to the C-terminal of the protein. Degenerate oligo nucleotide probes corresponding to this region and probes corresponding to the poly-A tail have been used as primers for a polymerase chain reaction to generate cytochrome P-450-specific probes from mRNA isolated from tissues known to be enriched in the enzyme of interest. Subsequently, these probes have been used for isolation of cDNA clones. There are some examples where this approach has proven successful, *e.g.* the isolation of the gene encoding the flavonoid 3'5'-hydroxylase involved in blue colour formation in petunia (Holton *et al.*, 1993). Often, however, this approach leads to isolation of a large number of cDNA clones encoding cytochromes P-450 of which the function is not known.

To characterize the biosynthetic enzyme system involved in the biosynthesis of the cyanogenic glucoside dhurrin, the classical approach of isolating the enzymes involved was taken. In the present review we will specifically comment on the isolation of cytochrome P-450$_{TYR}$, the first enzyme in the biosynthetic pathway for dhurrin. For isolation of this cytochrome P-450 enzyme, a new procedure based on dye column chromatography was developed (Fig. 2) (Sibbesen *et al.*, 1994). Micro-somal preparations from etiolated sorghum seedlings were used as starting material. The isolation procedure was designed to avoid high concentrations of the cytochrome P-450 enzymes, since this often leads to irreversible aggregation. All buffers were degassed and flushed with argon thrice prior to addition of detergent and dithiothreitol. The purification procedure was initiated by solubilization of the membrane bound enzymes from the microsomal preparation with a mixture of the non-ionic detergents RENEX 690 and Reduced TRITON X-100 followed by an ultracentrifugation step at 200 000 × g. The supernatant containing the solubilized enzymes was applied to a column containing a mixture of DEAE Sepharose fast flow and Sephacryl S-100 (20:80). The cytochromes P-450 were eluted with a CHAPS containing buffer under conditions where the NADPH-cytochrome P450 oxidoreductase and cytochrome b_5 remained bound to the column. These proteins were subsequently eluted from the column using a KCl-gradient. A homogenous NADPH-cytochrome P-450 oxidoreductase was obtained by affinity chromatography on a 2,5-ADP Sepharose column whereas cytochrome b_5 was further purified using a hydroxyapatite column. The cytochromes P-450 eluted from the ion exchange column in CHAPS-containing buffer were applied to a Cibachron Blue Agarose, from which cytochrome P-450$_{TYR}$ was eluted with a KCl-gradient. As a final step the cytochrome P-450$_{TYR}$ was applied to a Reactive Red 120

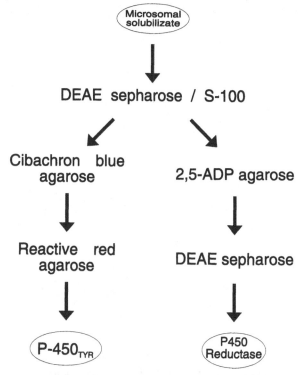

Fig. 2. Purification scheme for cytochrome P-450$_{TYR}$ and NADPH-cytochrome P-450 oxidoreductase involved in the biosynthesis of the cyanogenic glucoside dhurrin.

agarose column from which a highly purified cytochrome P-450$_{TYR}$ was eluted with a KCl gradient (Fig.3). During fractionation the total cytochrome P450 content of the individual fractions was monitored by carbon monoxide difference spectrum. The presence of the specific cytochrome P-450$_{TYR}$ was monitored by substrate binding spectra (Fig.4). For production of polyclonal antibodies and determination of *N*-terminal and internal amino acid sequences, cytochrome P-450$_{TYR}$ was further purified by preparative SDS–PAGE.

The molecular mass of cytochrome P-450$_{TYR}$ was found to be 57 kD as determined by SDS-PAGE. Polyclonal antibodies towards cytochrome P-450$_{TYR}$ and degenerated oligonucleotides derived from amino acid sequences of cytochrome P-450$_{TYR}$ were subsequently used to screen a cDNA library for a full length cDNA clone encoding cytochrome P-450$_{TYR}$ (Koch *et al.*, unpubl. results). This represents one of the first

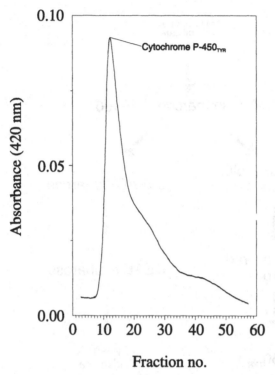

Fig. 3. Elution profile of the Cibachrome blue column.

isolated plant cytochrome P-450 with a known function, other than the ubiquitous cinnamic acid hydroxylases.

Characterization of cytochrome P-450$_{TYR}$

Reconstitution of cytochrome P-450 dependent monooxygenases has generally been achieved by insertion of purified cytochrome P-450 and NADPH-cytochrome P-450 reductase into artificial lipid vesicles (Ingelman-Sundberg, 1986). When L-tyrosine was administered to a reconstituted enzyme complex consisting of purified cytochrome P-450$_{TYR}$, NADPH-cytochrome P-450 oxidoreductase and L-α-dilauroylphosphatidylcholine in the presence of NADPH, p-hydroxyphenylacetaldehyde oxime accumulated in the reaction mixture (Sibbesen *et al.*, unpubl. results). This indicated that the first cytochrome P-450 monooxygenase is a multifunctional enzyme, which catalyses two consecutive N-hydroxylation reactions and a subsequent decarboxylation reaction

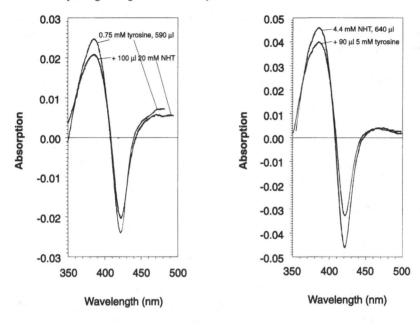

Fig. 4. Substrate binding spectra of P-450$_{TYR}$.

resulting in the overall conversion of L-tyrosine into *p*-hydroxyphe-nylacetaldehyde oxime. Quantitative substrate binding spectra substantiated these results: addition of *N*-hydroxytyrosine to cytochrome P-450$_{TYR}$ saturated with L-tyrosine did not alter the absorbance difference ($\Delta A_{390-420}$) in the substrate binding spectrum (Fig.4). Similarly, addition of L-tyrosine to cytochrome P-450$_{TYR}$ saturated with *N*-hydroxytyrosine did not increase the absorbance difference. These data demonstrated, that L-tyrosine and *N*-hydroxytyrosine are mutually exclusive and competing for the same binding site on the enzyme. In addition, heterologous expression in *Escherichia coli* of the isolated full length cDNA clone encoding cytochrome P-450$_{TYR}$ confirmed, that this single gene encodes a cytochrome P-450 enzyme, which is able to convert L-tyrosine into *p*-hydroxyphenylacetaldehyde oxime (Halkier *et al.*, unpubl.results). This surprising property of cytochrome P-450$_{TYR}$ explains the results obtained in previous experiments carried out using $^{18}O_2$, where the stereospecificity of the conversion of *N*-hydroxytyrosine into *p*-hydroxyphenylacetaldehyde oxime was analysed. In these studies, the microsomal enzyme system was documented to clearly distinguish between the two oxygen atoms introduced in the two consecutive

N-hydroxylation reactions (Halkier, Lykkesfeldt & Møller, 1991). This was difficult to reconcile with the known chemical properties of the suggested intermediates. During the diffusion of an intermediate from one active site to the other, free rotation around the C-N bond would be expected to render the two oxygen atoms indistinguishable in the subsequent enzymatic transformations. With the present knowledge, that the intermediates are kept within a single catalytic site, the results of these experiments become understandable. However, this questions the true nature of the intermediates identified between the amino acid and the oxime. Are these real intermediates? Would it be more correct to refer to these compounds as transition states? Or could some of the detected components represent artificially generated stable forms of these transition states, which could act as precursors for the microsomal enzyme system? These questions need to be addressed in carefully designed reconstitution experiments. Whether the nitro compound is a true intermediate is especially questionable, considering the reported very low metabolic production and conversion rate of 1-nitro-2-(*p*-hydroxyphenyl)ethane (Halkier & Møller, 1991).

In conclusion, cytochrome P-450$_{TYR}$ catalysing the first reactions in the biosynthesis of the cyanogenic glucoside dhurrin in *Sorghum bicolor* is multifunctional. Multifunctional cytochrome P-450 enzymes have to our knowledge not previously been purified and characterized in higher plants. In animal steroid biosynthesis there are at least four examples of purified cytochrome P-450 enzymes, which catalyse two or three subsequent and coordinated oxygen activations and oxygen insertions reactions at their active sites (Kühn-Veltend, Bunse & Förster, 1991).

Is glucosinolate biosynthesis related to the biosynthetic pathway of cyanogenic glucosides?

Glucosinolates (mustard oil glucosides) are a group of secondary plant products which upon hydrolysis forms a number of degradation products (isothiocyanates, thiocyanates, nitriles and oxazolidin-2-thiones), some of which possess potent biological activity (see the following two chapters). The presence of glucosinolates in *Brassicaceae* is of particular economic interest since many agricultural crops and important vegetables like oil seed rape, cabbages and broccoli are found in this family. Like cyanogenic glucosides, the glucosinolates are derived from amino acids (Kutacek, Prochazka & Veres, 1962). In addition, glucosinolates may be derived from precursor amino acids, which have undergone a series of chain elongations prior to the first *N*-hydroxylation reaction (Underhill, 1980). Administration of radioactively labelled compounds, presumed to be precursors, to whole plants have indicated, that the biosynthetic pathway leading to glucosinolates may involve N-hydroxyamino acids, oximes, and nitro compounds as intermediates. Exactly the same intermediates which have been isolated and identified during the elucidation of the biosynthetic pathway leading to cyanogenic glucosides (Underhill, 1967; Kindl & Underhill, 1968; Matsuo, Kirkland & Underhill, 1972). This led to the suggestion, that the initial reactions in the biosynthesis of these two groups of secondary plant products are identical (Møller, 1981). As described above, the conversion of tyrosine to p-hydroxyphenylacetaldehyde oxime in the sorghum system is carried out by one multifunctional cytochrome P-450. It is tempting to speculate, that the equivalent reactions in the biosynthetic pathway leading to glucosinolates are carried out by a homologous cytochrome P-450. The aldehyde oxime might then be the diverging point for the biosynthesis of glucosinolates and cyanogenic glucosides, as first suggested by Mahadevan in 1975 (Mahadevan, 1975).

Recently, it was demonstrated that a microsomal preparation from oil seed rape leaves was able to convert homophenylalanine (Dawson *et al.*, 1993) and homomethionine (Bennett *et al.*, 1993) to the corresponding aldehyde oximes. Both conversions showed NADPH dependency. Characterization of the enzyme systems have led to the conclusion that these two enzyme systems are not dependent on cytochrome P-450. Both conversions were suggested to be catalyzed by flavin-containing monooxygenases (Bennett *et al.*, 1993). *Tropaeolum majus* contains only a single glucosinolate, tropaeolin, which is derived from the aromatic amino acid phenylalanine. Using phenylalanine as substrate, Bennett and Wallsgrove (pers. comm.) have succeeded in detecting an NADPH-dependent decarboxylation reaction in a microsomal preparation from *Tropaeolum majus*. However, the required buffer composition for the isolation of a biosynthetically active microsomal preparation were different for *Tropaeolum majus* than for oil seed rape preparations (Bennett & Wallsgrove, pers.comm.). The amino acid metabolizing enzyme system in *Tropaeolum majus* may therefore differ from the enzyme system in rape. We have observed a cross reaction of polyclonal antibodies raised against cytochrome P-450$_{TYR}$ from sorghum with a 57 kD protein in microsomal preparations of *Tropaeolum majus* (Rouzé, unpubl.results). In addition, the cDNA clone encoding cytochrome P-450$_{TYR}$ in sorghum hybridizes to genomic DNA from *Tropaeolum majus* (Halkier, unpubl.results). This suggests, that the enzyme catalysing the N-hydroxylation of phenylalanine in the biosynthesis of tropaeolin in *Tropaeolum majus* could be homologous to cytochrome P-450$_{TYR}$. The unravelling of the relationship between the

238   O. SIBBESEN *et al.*

biosynthesis of cyanogenic glucosides and glucosinolates seems to lie close ahead.

Application of biotechnology

Cytochrome P-450$_{TYR}$ catalyses the first and rate-limiting reaction in the biosynthesis of cyanogenic glucosides. In addition, the high substrate specificity of this enzyme determines which cyanogenic glucoside is made. The available evidence suggests that the homologous enzymes present in other cyanogenic plants and possible also in glucosinolate producing plants will possess the same characteristics. The isolation of the gene encoding this enzyme provides a possible tool for modulation of the level of cyanogenic glucosides in plants. Improved nutritional value of agriculturally important crops could be obtained by tissue-specific elimination of cyanogenic glucosides by blocking the biosynthetic pathway using antisense techniques. Tissue-specific changes of the glucosinolate profile could be designed to improve nutritional value of a part of a plant. In the biosynthesis of glucosinolates, the thiohydroximate glucosyltransferase and the sulphotransferase positioned after the initial *N*-hydroxylations of the amino acid have little specificity for the varied side chain structures (Reed *et al.*, 1993). The composition of glucosinolates in specific tissue could thus be controlled by alteration of the substrate-specificity of the enzyme initiating the biosynthesis by site-directed mutagenesis.

To date, three bacterial cytochromes P-450 have been crystallized. One of these, cytochrome P-450$_{BM-3}$, resembles the microsomal cytochromes P-450 (Ravichandran *et al.*, 1993). The cytochrome P-450$_{BM-3}$ structure, combined with already available data from site-directed mutagenesis studies, provide qualified predictions as to which amino acids are important in determining substrate-specificity. This can prove a very valuable tool for design of cyanogenic glucoside and glucosinolate profiles in plants, based on manipulation of the substrate-specificity of the oxime-generating enzymes.

References

Akazawa, T., Miljanich, P. & Conn, E.E. (1960). Studies on cyanogenic glucosides of *Sorghum vulgare*. *Plant Physiology*, **35**, 535–8.
Bellotti, A.C. & Arias, B. (1993). The possible role of HCN on the biology and feeding behaviour of the cassava burrowing bug (*Cyrtomenus bergi* Froeschner). In: *Proceedings of the First International Scientific Meeting of the Cassava Biotechnology Network*, ed. W.M.Roca & A.M.Thro, pp. 406–9. Cali, Columbia: CIAT.

Bennett, R.N., Donald, A., Dawson, G., Hick, A. & Wallsgrove, R. (1993). Aldoxime-forming microsomal enzyme systems involved in the biosynthesis of glucosinolates in oilseed rape (*Brassica napus*) leaves. *Plant Physiology*, **102**, 1307–12.

Collinge, D.B. & Hughes, M.A. (1982). *In vitro* characterization of the *Ac*-Locus in white clover (*Trifolium repens*). *Archives of Biochemistry and Biophysics*, **218**, 38–45.

Cook, R., McCaig, N., McMillan, J.M.B. & Lumsden, W.B. (1990). Ethyl carbamate formation in grain-based spirits. Part III. The primary source. *Journal of the Institute of Brewing*, **96**, 233–44.

Cutler, A.J. & Conn, E.E. (1981). The biosynthesis of cyanogenic glucosides in *Linum usitatissimum* (linen flax) *in vitro*. *Archives of Biochemistry and Biophysics*, **212**, 468–74.

Dawson, G.W., Hick, A.J., Bennett, R.N., Donald, A., Pickett, J.A. & Wallsgrove, R.M. (1993). Synthesis of glucosinolate precursors and investigations into the biosynthesis into phenylalkyl- and methylthioalkylglucosinolates. *Journal of Biological Chemistry*, **268**, 27154–9.

Donaldson, R.P. & Luster, D.G. (1991). Multiple forms of plant cytochromes P-450. *Plant Physiology*, **96**, 669–74.

Guengerich, P. (1991). Reactions and significance of cytochrome P450 enzymes. *Journal of Biological Chemistry*, **266**, 10019–22.

Halkier, B.A., Lykkesfeldt, J. & Møller, B.L. (1991). 2-Nitro-3-(*p*-hydroxyphenyl)propionate and aci-nitro-2-(p-hydroxyphenyl)-ethane, two intermediates in the biosynthesis of the cyanogenic glucoside dhurrin in *Sorghum bicolor* (L.) Moench. *Proceedings of the National Academy of Sciences, USA*, **88**, 487–91.

Halkier, B.A. & Møller, B.L. (1990). The biosynthesis of cyanogenic glucosides in higher plants: identification of three hydroxylation steps in the biosynthesis of dhurrin in *Sorghum bicolor* (L.) Moench and the involvement of the 1-*aci*-nitro-2-(*p*-hydroxyphenyl)ethane as an intermediate. *Journal of Biological Chemistry*, **265**, 21114–21.

Halkier, B.A. & Møller, B.L. (1991). Involvement of cytochrome P-450 in the biosynthesis of dhurrin in *Sorghum bicolor* (L.) Moench. *Plant Physiology*, **96**, 10–7.

Halkier, B.A., Olsen, C.E. & Møller, B.L. (1989). The biosynthesis of cyanogenic glucosides in higher plants: the (*E*)- and (*Z*)-isomers of *p*-hydroxyphenylacetaldehyde oxime as intermediates in the biosynthesis of dhurrin in *Sorghum bicolor* (L.) Moench. *Journal of Biological Chemistry*, **264**, 19487–94.

Hösel, W. & Nahrstedt, A. (1980). *In vitro* biosynthesis of the cyanogenic glucoside taxiphyllin in *Triglochin maritima*. *Archives of Biochemistry and Biophysics*, **203**, 753–7.

Holton, T.A., Brugliera, F., Lester, D.R., Tanaka, Y., Hyland, C.D., Menting, J.G.T., Lu, C.Y., Farcy, E., Stevenson, T.W. & Cornish, E.C. (1993). Cloning and expression of cytochrome-P450 genes controlling flower colour. *Nature, London*, **366**, 276–9.

240   O. SIBBESEN *et al.*

Ingelman-Sundberg, M. (1986). Cytochrome P-450 organization and membrane interactions. In *Cytochrome P450. Structure, Mechanism and Biochemistry*, ed. P.R.Ortiz de Montellano, pp. 119–60: Plenum Press.

Kindl, H. & Underhill, E.W. (1968). Biosynthesis of mustard oil glucosides: N-hydroxyphenylalanine, a precursor to glucotropaeolum and a substrate for the enzymatic and non-enzymatic formation of phenylacetaldehyde aldoxime. *Phytochemistry*, 7, 745–56.

Koch, B., Nielsen, V.S., Halkier, B.A., Olsen, C.E. & Møller, B.L. (1992). The biosynthesis of cyanogenic glucosides in seedlings of cassava (*Manihot esculenta* Crantz). *Archives of Biochemistry and Biophysics*, 292, 141–50.

Kühn-Veltend, W.N., Bunse, T. & Förster, M.E.C. (1991). Enzyme kinetic and inhibition analyses of cytochrome P450XVII, a protein with a bifunctional catalytic site. *Journal of Biological Chemistry*, 266, 6291–301.

Kutacek, M., Prochazka, Z. & Veres, K. (1962). Biogenesis of oximes in nitrogen metabolism in plants. *Nature*, 194, 393–4.

Lieberei, R. (1986). Cyanogenesis of *Hevea brasiliensis* during infection with *Microcyclus ulei*. *Journal of Phytopathology*, 115, 134–46.

McFarlane, I.J., Lees, E.M. & Conn, E.E. (1975). The *in vitro* biosynthesis of dhurrin, the cyanogenic glucoside of *Sorghum bicolor*. *Journal of Biological Chemistry*, 250, 4708–13.

Mahadevan, S. (1975). The roles of oximes in nitrogen metabolism in plants. *Annual Review of Plant Physiology*, 24, 69–88.

Matsuo, M., Kirkland, D.F. & Underhill, E.W. (1972). 1-Nitro-2-phenylethane, a possible intermediate in the biosynthesis of benzyl-glucosinolate. *Phytochemistry*, 11, 697–701.

Møller, B.L. (1981). The involvement of N-hydroxyamino acids as intermediates in metabolic transformation. In *Cyanide in Biology*, ed. B.Vennesland, E.E.Conn, C.J.Knowles, J.Westley & F.Wissing, pp. 197–215. London & New York: Academic Press.

Møller, B.L. & Conn, E.E. (1979). The biosynthesis of cyanogenic glucosides in higher plants: N-hydroxytyrosine as an intermediate in the biosynthesis of dhurrin by *Sorghum bicolor*. *Journal of Biological Chemistry*, 254, 8575–83.

Møller, B.L. & Conn, E.E. (1980). The biosynthesis of cyanogenic glucosides in higher plants: channelling of intermediates in dhurrin biosynthesis by a microsomal system from *Sorghum bicolor* (linn) Moench. *Journal of Biological Chemistry*, 255, 3049–56.

Møller, B.L. & Poulton, J.E. (1993). Cyanogenic glucosides. In *Methods of Plant Biochemistry, 9: Enzymes of Secondary Metabolism*, ed. P.J.Lea, pp. 183–207. London: Academic Press.

Pourmohseni, H. & Ibenthal, W.D. (1991). Novel β-cyanoglucosides in the epidermal tissue of barley and their possible role in the barley-powdery mildew interaction. *Angewandte Botanik* 65, 341–50.

Ravichandran, K.G., Boddupalli, S.S., Hasemann, C.A., Peterson,

J.A. & Deisenhofer, J. (1993). Crystal structure of hemoprotein domain of P450BM-3, a prototype for microsomal P450s. *Science,* **261**, 731–6.

Reed, D.W., Davin, L., Jain, J.C., Deluza, V., Nelson, L. & Underhill, E.W. (1993). Purification and properties of UDP-glucose: thiohydroximate glucosyltransferase from *Brassica napus* L. seedlings. *Archives of Biochemistry and Biophysics,* **305**, 526–32.

Shimada, M. & Conn, E.E. (1977). The enzymatic conversion of *p*-hydroxyphenylacetaldoxime to *p*-hydroxymandelonitrile. *Archives of Biochemistry and Biophysics,* **180**, 199–207.

Sibbesen, O., Koch, B., Halkier, B.A. & Møller, B.L. (1994). Isolation of the heme-thiolate enzyme cytochrome P-450$_{tyr}$ which catalyses the committed step in the biosynthesis of the cyanogenic glucoside dhurrin in *Sorghum bicolor* L. Moench. *Proceedings of the National Academy of Sciences, USA* (in press).

Underhill, E.W. (1967). Biosynthesis of mustard oil glucosides: conversion of phenylacetaldehyde oxime and 3-phenylpropionaldehyde oxime to glucotropaeolin and gluconasturtin. *European Journal of Biochemistry,* **2**, 61–3.

Underhill, E.W. (1980). Glucosinolates. In *Encyclopedia of Plant Physiology New Series*, Vol. 8, ed. E.A. Bell & B.V. Charlwood, pp. 493–511. Berlin: Springer Verlag.

ROGER M. WALLSGROVE and
RICHARD N. BENNETT

The biosynthesis of glucosinolates in *Brassicas*

Introduction

The glucosinolates are a group of N- and S-containing secondary metabolites, found in several plant families (Rodman, 1981) of which the most significant from an agricultural and culinary viewpoint are the *Brassicaceae*. Glucosinolates have the common structure shown in Fig.1, comprising three main elements: the R-C-N structure derived from an amino acid, a glucose moiety attached via a thioester link, and the sulphate group linked to the nitrogen. To date, about 100 different glucosinolates have been identified in plants, derived from a variety of amino acid precursors. Glucosinolate-containing plants also possess thioglucosidases, 'myrosinases' (EC 3.2.3.1), which hydrolyse glucosino-

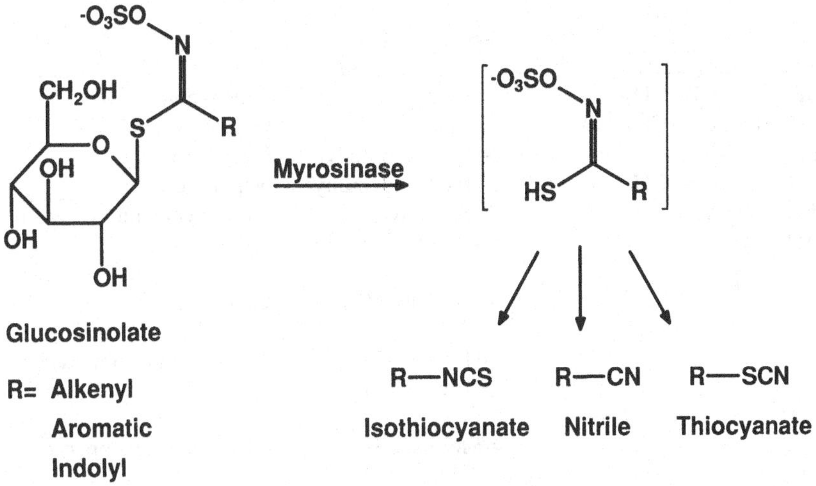

Fig. 1. The structure of glucosinolates, and their myrosinase-catalysed breakdown products.

lates. Following cleavage of the thioester link to release glucose, the resulting intermediate undergoes spontaneous rearrangement to a variety of products, including isothiocyanates, thiocyanates, and nitriles (Fig.1), the product(s) depending on the pH of the medium and other factors. These breakdown products are the flavour components of mustards, horseradish, and other *Brassica* crops, and they have fungitoxic, antibacterial, and other defence-related properties (for review see Chew, 1988). In the intact plant, glucosinolates and myrosinases are probably physically separated (the details of their precise localization are a matter of debate, see discussion in Poulton & Møller, 1993), but they come together to react when tissues are disrupted.

In *Brassica* species, especially the agriculturally important *B.napus, B.campestris*, and related species, a limited range of glucosinolates are found (Table 1). The glucosinolates, like the cyanogenic glucosides, are derived from amino acids. There are three main structural types of glucosinolates in Brassicas: aromatic glucosinolates derived from phenylalanine (and possibly tyrosine), aliphatic and alkenyl glucosinolates derived from methionine, and indole glucosinolates derived from tryptophan (Sorensen, 1991). Different tissues, at different stages of development, have quite different glucosinolate profiles (Fieldsend &

Table 1. *The major glucosinolates found in* Brassica napus

R-group structure	Semi-systematic name (-glucosinolate)	Trivial name
$CH_2=CH-CH_2-CH_2-$	Butenyl	Gluconapin
$CH_2=CH-CH_2-CH_2-CH_2$	Pentenyl	Glucobrassicanapin
$CH_2=CH-CH-CH_2-OH$	2-Hydroxybutenyl	Progoitrin
$CH_2=CH-CH_2-CH-CH_2-OH$	2-Hydroxypentenyl	Napoleiferin
Phe-CH_2-CH_2-	Phenylethyl	Gluconasturtiin
HO-**Phe**-CH_2-	p-Hydroxybenzyl	Sinalbin
(indol-3-yl) CH_2-	3-Indolylmethyl	Glucobrassicin
(4-hydroxyindol-3-yl) CH_2-	4-Hydroxy-3-indolylmethyl	4-Hydroxyglucobrassicin
(4-methoxyindol-3-yl) CH_2-	4-Methoxy-3-indolylmethyl	4-Methoxyglucobrassicin

Milford, 1994). New glucosinolates are still being discovered in Cruciferous plants each year, and structures derived from other amino acid precursors are found in other species. Fenwick, Heaney & Mullin, (1983) have reviewed the identity and localisation of some of the glucosinolates found in food plants.

In many cases, particularly with the methionine-derived compounds, structures derived from the protein amino acid are not found, but instead there appears to be a side-chain elongation (discussed in more detail in the next chapter), which takes place before the amino acids enter the biosynthetic pathway. In addition, a variety of modifications take place once the glucosinolates have been formed, such as side-chain hydroxylation or methoxylation, and loss of the methylthio group or oxidation of the methylthio-S in aliphatic and alkenyl glucosinolates.

The biosynthetic pathway for the conversion of amino acids to glucosinolates, as currently understood, is shown in Fig. 2. The various steps, intermediates, and enzymes involved, are discussed in detail below.

Conversion of the amino acid precursors to aldoximes

Until recently the biosynthesis of glucosinolates was poorly understood. Preliminary information on the biosynthetic pathway was deduced from feeding whole plants with various radiolabelled compounds (for reviews see, Underhill, Wetter & Chisholm, 1973; Kjaer & Larsen, 1973, 1976, 1977, 1980; Larsen, 1981). Much has been made of the similarities between the biosynthesis of cyanogenic glucosides and glucosinolates, particularly with regard to the potential intermediates and the nature of the enzymes involved (for review, see Poulton & Møller, 1993, and also discussion in the previous chapter). However, recent research has shown that, although there are some common intermediates in the two pathways, the enzymes that catalyse the first steps of glucosinolate and cyanogenic glucoside biosynthesis are different. The following section outlines what is known about the biosynthesis of glucosinolates, and where the pathway differs from cyanogenic glucoside biosynthesis.

The initial research on glucosinolate biosynthesis involved feeding ^{14}C-labelled precursors to whole plants or plant tissues. Kutachek, Prochazka & Veres (1962) fed DL-^{14}C-U-Trp to cabbages and demonstrated incorporation of the label into the indole glucosinolate glucobrassicin (indol-3-methyl glucosinolate). Underhill *et al.* (1962) and Benn (1962) showed that ^{14}C-L-Phe applied to *Tropaeolum majus* was incorporated into benzyl isothiocyanate. They concluded that the L-Phe was first converted into benzyl-glucosinolate (glucotropaeolin) and that the subsequent action of myrosinase, upon tissue disruption, led to

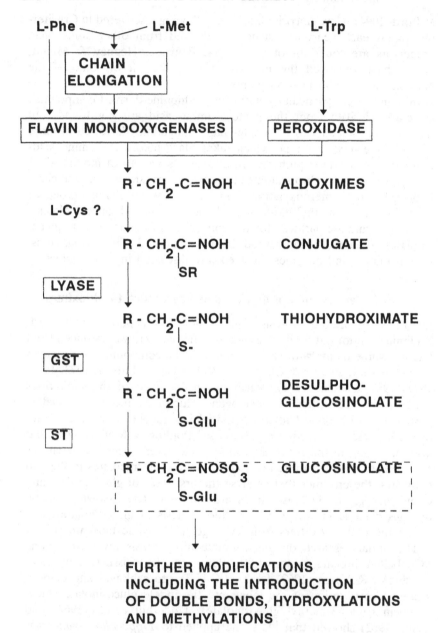

Fig. 2. The postulated biosynthetic pathway leading to glucosinolates. GST, glucosyltransferase; ST, sulphotransferase.

benzyl-isothiocyante formation. Underhill and Chisolm (1964) have shown that ^{14}C-labelled phenylethylamine, phenylacetamide, phenylace-tohydroxamic acid and phenylpyruvic acid aldoxime were not precursors or intermediates in benzyl-glucosinolate biosynthesis. They demon-strated, using dual labelled (^{14}C and ^{15}N) L-Phe, that the ^{15}N was retained in the glucosinolate that was formed, and that the intermediates in glucosinolate biosynthesis must therefore be nitrogenous. Incorpor-ation of ^{14}C-Trp into the indole glucosinolates glucobrassicin and neo-brassicin in hypocotyls of *Sinapis alba* (mustard) was demonstrated (Schraudolf & Bergmann, 1965); there was no detection of label in IAA (indole acetic acid) or IAN (indole acetonitrile). Application of ^{14}C-L-Phe and ^{14}C-sodium acetate to *Nasturtium officinale* (watercress) lead to accumulation of label in gluconasturtiin (phenylethyl glucosinolate) (Underhill, 1965).

Chisolm and Wetter (1966) investigated the biosynthesis of sinigrin (allyl glucosinolate) in *Amoracia lapathifolia* (horseradish). They found that feeding either ^{14}C-L-Met, ^{14}C-L-homoserine or ^{14}C-L-homo-methionine led to the incorporation of label in sinigrin; the incorporation was far greater when ^{14}C-L-homomethionine was used. They proposed that homoserine was first converted into cystathionine, then methionine, and chain-extended to produce homomethionine which was converted into sinigrin. As well as the glucosinolates derived directly from protein amino acids (L-Phe, L-Tyr and L-Trp) there are a large number derived from amino acid homologues. Common chain-extended glucosinolates include phenylethylglucosinolate (the precursor is homophenylalanine), but-3-enyl-glucosinolate (precursor is dihomomethionine) and pent-4-enyl-glucosinolate (precursor is trihomomethionine).

Tapper and Butler (1967) showed in a key paper that either ^{14}C-L-Phe or ^{14}C-L-phenylacetaldoxime were equally suitable precursors for benzyl-glucosinolate in *Lepidium sativum* (garden cress), and also that ^{14}C-isobutylaldoxime was a precursor of isopropylglucosinolate in *Cochleria officinalis*. When either ^{14}C-Phe or ^{14}C-phenylacetaldoxime were fed to *Tropaeolum majus* shoots label was found in benzyl-glucosinolate (Underhill, 1967). In a trapping experiment it was shown, for the first time, that ^{14}C-Phe fed to the plants was converted into ^{14}C-phenylacetaldoxime thus proving it was a true intermediate in the pathway. Underhill (1968) also showed that ^{14}C-2-amino-4-phenylbutyric acid and ^{14}C-phenylpropionaldehyde oxime were precursors of phenyle-thyl-glucosinolate in *Nasturtium officinale* (watercress). Conn (1981) has reported the wide spread occurrence of aldoximes in plants, which provides indirect evidence for their role in glucosinolate biosynthesis.

Several potential nitrogenous intermediates between the amino acid precursors and the aldoximes have been proposed, including N-hydroxy amino acids and oximino acids (Tapper & Butler, 1967; Underhill & Wetter, 1969). It was found that soluble enzyme preparations from *Sinapis alba*, *Tropaeolum majus*, and *Nasturtium officinale* were able to catalyse the conversion of ^{14}C-N-hydroxyphenylalanine to ^{14}C-phenylacetaldoxime, an intermediate in benzyl-glucosinolate biosynthesis (Kindl & Underhill, 1968). The enzyme catalysing this reaction had a pH optima of 8.6, was stimulated by FMN, required oxygen, was unaffected by either KCN or *p*-chloromercuriphenylsulfonate, but was inhibited by DTT. There was quite a high background of non-enzymic conversion, which could be stimulated by addition of Fe^{3+} ions and o-iodosobenzoate. It was also shown that ^{14}C-2-oximino-3-phenylpropionic acid (another potential intermediate) was not converted into phenylacetalaldoxime (Kindl & Underhill, 1968). Recent work using N-hydroxy-homophenylalanine (a potential intermediate in phenylethylglucosinolate biosynthesis) has shown that the N-hydroxy amino acid spontaneously, and non-enzymically, degrades into phenylethylaldoxime in pH 7.5 phosphate buffers (R.N.Bennett & A.J.Hick, unpublished observations). The non-enzymic degradation of other N-hydroxy amino acids has been demonstrated (Halkier, Olsen & Møller, 1989). There have been no reports of the natural occurrence of N-hydroxy amino acids in any plant species, and it is possible that the apparent activity with such compounds is an artefact.

Matsuo, Kirkland & Underhill (1972) proposed that nitro-2-phenylethanc was an intermediate in benzyl-glucosinolate biosynthesis in *Tropaeolum majus*. Feeding either ^{14}C-Phe or ^{14}C-nitro-2-phenylethane to the plants lead to label accumulating in benzylglucosinolate. A trapping experiment appeared to show that the nitro-2-phenylethane was a true intermediate. However, the reported details of the experiment were incomplete, and no other researchers have demonstrated that nitro-2-phenylethane is a true intermediate.

Ettlinger and Kjaer (1968) proposed that the α-nitroso acid of phenylalanine may be an intermediate in benzyl-glucosinolate biosynthesis. Since these initial studies numerous exotic nitrogenous intermediates have been proposed; most of these suggested by the research on cyanogenic glycoside biosynthesis (for a review, see Poulton & Møller, 1993), on the assumption that the two pathways were similar. With many of the early studies on the potential intermediates of glucosinolate biosynthesis, very few of the experiments have been repeated. Detailed dual-labelling experiments and/or trapping experiments irrefutably demonstrating the real glucosinolate intermediates have not been tried. The chemical nature of the glucosinolate precursors (amino acids and their homologues) and some of the intermediates (aldoximes and *aci*-nitro compounds) are now known, but few researchers have considered the enzymes involved in glucosinolate biosynthesis. Most of the assumptions about the glucosinolate pathway enzymes have been based on comparisons with cyanogenic glucoside biosynthesis. Until recently there have been no reports on the type of enzymes involved in the first steps of glucosinolate biosynthesis. The conversion of L-Trp to indoleacetaldoxime (the potential precursor of the indole glucosinolates and IAA) has been demonstrated in microsomal preparations from etiolated seedlings of *Brassica campestris* (Ludwig-Muller & Hilgenberg, 1988). This activity was also detected in hypocotyl and seedling extracts from maize, sunflower, tobacco and pea (Ludwig-Muller & Hilgenberg, 1988). The activity appeared to be due to several peroxidases with pIs of 8–9 (Ludwig-Muller *et al.*, 1990). It was proposed that the enzymes may be important in the biosynthesis of IAA, IAN, and possibly the indole glucosinolates.

It has been assumed that the enzymes catalysing the formation of aldoximes in oilseed rape and other Brassica species are cytochrome P450s, as for the cyanogenic glucosides (see previous chapter). In all of the earlier attempts to detect aldoxime-forming activities in Brassica species there were two major problems. Etiolated seedlings were used, and the extraction and assay conditions had been optimized for cytochrome P450-type enzymes.

In our laboratory we have found that microsomal preparations from young green leaves of *Brassica napus* (oilseed rape) will catalyse the NADPH-dependent decarboxylation of homophenylalanine (the precursor for phenylethylglucosinolate) and dihomomethionine (the precursor for the but-3-enylglucosinolates), to their respective aldoximes (Bennett *et al.*, 1993; Dawson *et al.*, 1993). The oilseed rape microsomal enzymes have very tight substrate specificities (there is clear evidence for at least two separate activities, specific for aliphatic and aromatic amino acids, respectively), pH optimas of 7.5, are unaffected by haem or cytochrome P450 inhibitors, and are unaffected by N-ethylmaleimide (NEM) or phenylmethylsulfonylfluoride (PMSF). The activities are strongly inhibited by DTT, copper salts and diethylpyrocarbonate. These characteristics are very similar to the properties of the flavin-containing monooxygenases (FMOs), involved in xenobiotic detoxification, that are found in mammalian systems (Ziegler, 1988; Coecke *et al.*, 1992). The insensitivity of the enzymes to haem inhibitors, cytochrome P450 inhibitors and NEM, and also their sensitivity to DTT are particularly characteristic of FMOs (Coeke *et al.*, 1992). The mammalian enzymes catalyse the conversion of secondary and tertiary amines to aldoximes, and also the oxidation of various sulphur-containing compounds (Ziegler, 1988).

The activities of the two oilseed rape FMOs are greatest in young expanding leaves, the time at which glucosinolate accumulation is known to be high (Porter et al., 1991), and decrease rapidly as the leaf matures. Unlike the channelled intermediate system involved in the biosynthesis of the cyanogenic glucoside dhurrin in Sorghum (Halkier et al., 1989), it appears that some of the glucosinolate intermediates are freely exchangeable with the medium (Dawson et al., 1993). The appearance of the activities of the two oilseed rape FMOs is apparently light dependent, and restricted to true leaves of oilseed rape. Neither FMO activity was detected in etiolated or light grown cotyledons of oilseed rape (tissues in which aliphatic and aromatic glucosinolates do not accumulate), and in light- or nutrient-stressed (SO_4^{2-} and NO_3^-) plants the foliar activity was significantly reduced (Kiddle, Bennett & Wallsgrove, unpublished observations).

The biosynthesis of benzyl-glucosinolate in Tropaeolum majus raises several questions about glucosinolate biosynthesis in other Crucifers. Lykkesfeldt and Møller (1993) found that ^{14}C-L-Phe was incorporated into benzylglucosinolate when fed to T.majus seedlings and mature plants. They tried to detect a channelled microsomal system, similar to that found in Sorghum, that would convert L-Phe to benzyl-glucosinolate. No such activity was detected, and they proposed that isothiocyanate formation inhibited the in vitro biosynthesis of benzyl-glucosinolate (Lykkesfeldt & Møller, 1993). There are two major problems associated with assaying the complete glucosinolate biosynthetic pathway. First the optimization of all of the enzymes in the pathway needs to be achieved (it is not known how many enzymes are involved). Secondly the last two steps of the pathway are catalysed by cytosolic and not microsomal enzymes (see following sections). We have found a cytochrome P450-type enzyme in microsomes from very young leaves of T.majus that is NADPH-dependent, has a pH optima of 7.5, and oxidatively decarboxylates L-Phe; the activity was very weak by comparison with the rape leaf FMO activities. This enzyme may be very sensitive to benzyl isothiocyanate (isothiocyanates are biologically very reactive) and this would partly explain why no benzylglucosinolate biosynthesis was detected in the microsomal preparations from T.majus.

It is thus possible that the mechanisms of aldoxime formation in Crucifers and cyanogenic plants have evolved separately, despite the apparent similarity in respect of the amino acid precursors and some of the intermediates. Care should therefore be taken when assigning any taxonomic importance to the presence of compounds such as aldoximes, which are known to be distributed throughout the plant kingdom (Conn, 1981). If a cytochrome P450-mediated aldoxime formation can be confirmed in *Tropaeolum* (and FMO-like activity ruled out), it would seem that glucosinolate biosynthesis itself may have evolved independently at least twice. Taxonomic relationships based largely on the presence of glucosinolates would then look to be very unreliable.

The formation of thiohydroximates from aldoximes

The biosynthesis of thiohydroximates in oilseed rape and other Crucifers is the least understood aspect of glucosinolate biosynthesis. It has been suggested that the aldoxime (or possibly the *aci*-nitro compound) is conjugated to cysteine or possibly methionine (for reviews, see Underhill *et al.*, 1973; Poulton & Møller, 1993). When *Tropaeolum majus* shoots were fed phenyl(^{14}C-1)acetaldoxime it was possible to detect formation of radiolabelled phenyl-acetothiohydroximate (Underhill & Wetter, 1969). There is some evidence suggesting that the source of the thioglucosidic sulphur varies depending on the plant species (Kindl, 1964, 1965; Wetter & Chisolm, 1968). The aldoxime–cysteine conjugate is thought to be cleaved by a CS-lyase to produce the thiohydroximate.

CS-lyases are ubiquitous in plants and animals. In animals they are associated with glutathione-S-transferases involved in xenobiotic detoxification (for reviews, see Mannervik & Danielson, 1988; Pickett & Lu, 1989). Xenobiotics are oxidized (by either cytochrome P450-type enzymes or FMOs), conjugated to glutathione, converted to cysteine conjugates, then cleaved by a CS-lyase (e.g. a β-lyase) to thiols, pyruvic acid and ammonia. The thiols are usually deactivated by conjugation to glucose (from UDP-glucose). Glutathione-S-transferases have also been detected in plants and may be involved in herbicide detoxification (Timmerman, 1989). CS-lyases have also been found in plants (Mazelis, Scott & Gallie, 1982; Hall & Smith, 1983; Hamamato & Mazelis, 1986; Staton & Mazelis, 1991). The most common plant CS-lyase is cystathionase, which catalyses the cleavage of cystathionine to homocysteine, in the methionine biosynthetic pathway (Umbarger, 1978; Giovanelli, Mudd & Dakto, 1980; Wallsgrove, Lea & Miflin, 1983). It is therefore possible to envisage a pathway in Crucifers leading to thiohydroximate formation involving a cysteine conjugate and a CS-lyase. Cysteine would be conjugated to the aldoxime, and this conjugate cleaved by a β-lyase to yield the thiohydroximate, pyruvic acid and ammonia.

As yet no conjugated intermediates or enzymes that catalyse thiohydroximate formation have been characterised in any Crucifer. There is obviously a considerable amount of work to be done on these early steps in glucosinolate biosynthesis (from amino acid precursors to the thiohydroximates) before the intermediates and enzymology are fully established. There is the question of whether the biosynthesis of glucosinolates in different Crucifer species involves the same intermediates, and whether the enzymes that catalyse the formation of these intermediates are common to all Crucifers. Evidence suggests that this may not be the case. The only common aspects of the individual pathways may be the use of amino acids as precursors, and the formation of aldoximes and thiohydroximates.

There is evidence to suggest that thiohydroximate formation from aldoximes involves enzymes with broad specificity, as addition of synthetic aldoximes to Brassica seedlings or tissue cultures leads to the production of the appropriate artificial glucosinolates (Grootwassink, Balsevich & Kolenovsky, 1990). Other compounds are produced from these aldoximes, particularly glucosides and sulphatoglucosides, and the ability to carry out such metabolic conversions of aldoximes is found in many plant species, not simply those which make glucosinolates (Grootwassink et al. 1990). Aldoximes are probably ubiquitous in plants (Conn, 1981), acting as precursors to many metabolites, including the hormone IAA (Helmlinger, Rausch & Hilgenberg, 1987).

Glycosylation and sulphation

Conversion of thiohydroximates to glucosinolates proceeds via the desulphoglucosinolates, formed by glycosylation of thiohydroximates. This step, and the subsequent sulphation to produce the final glucosinolate, have been studied in more detail than any other part of the biosynthetic pathway, and the enzymes characterised in some detail. Their extraction, assay, and characterisation have been reviewed recently (Poulton & Møller, 1993).

The glycosylation reaction is catalysed by a UDPG-dependent enzyme, UDPG:thiohydroximate glucosyltransferase (EC 2.4.1.-). This enzyme has been partially purified from both Tropaeolum majus and Brassica juncea, and detected in a variety of other glucosinolate-producing plants. It appears to be a soluble cytosolic protein, with a molecular mass of 44–40 KDa. Activity is specific for thiohydroximates, but a wide range of such compounds act as substrates and there is no evidence for isozymes with different substrate specificities.

At least in B.juncea, the glucosyltransferase is closely associated with the next enzyme, which catalyses the PAPS-dependent formation of glucosinolates from desulphoglucosinolates (PAPS: desulphoglucosinolate sulphotransferase, EC 2.8.2.-) (Jain *et al.*, 1990). The enzyme is very unstable, which has hindered attempts to purify it. The link with the glucosyltransferase is intriguing, and the failure to separate the two activities with a variety of protein separation techniques suggests the two enzyme activities are either located on the same polypeptide, or consist of two polypeptides in a functional complex.

The sulphotransferase from *Lepidium sativum* (cress) was shown to be active with both desulphobenzylglucosinolate and desulphoallylglucosinolate (Glendening & Poulton, 1988), suggesting little or no side chain specificity for this enzyme. A variety of other, non-glucosinolate related, substrates were tested, but the partially-purified enzyme did not sulphate any of them (Glendening & Poulton, 1990).

As these last two enzymes in the main biosynthetic pathway are soluble, cytosolic proteins, it is clear that the channelled, membrane-bound complex found in cyanogenic glucoside biosynthesis is not mirrored in glucosinolate biosynthesis. At least this is so in Brassica species, the nature of the biosynthetic machinery in non-Crucifer species may be different, and needs to be investigated.

Side-chain modifications

Aliphatic, aromatic and indolyl glucosinolates all undergo side chain modification, which is assumed to occur after the parent glucosinolate has been synthesized. Such modifications can include hydroxylation and methoxylation, and for the aliphatic glucosinolates removal of the terminal methythiol to produce alkenyl compounds. Little is known of the biochemistry of such modifications, but a cytochrome P-450 linked hydroxylase has been described (Rossiter, James & Atkins, 1990) which acts on butenyl (and possibly pentenyl) glucosinolates. The order in which various side-chain modifications occur is uncertain, especially as the sulphotransferase enzyme will apparently accept unsaturated alkenyl glucosinolate precursors (Glendening & Poulton, 1988), but the hydroxylase appears to act on glucosinolates *per se* rather than any earlier intermediate. Some aspects of the genetics of glucosinolate side-chain modification are discussed in the following chapter.

Developmental and environmental influences on glucosinolate biosynthesis

In unstressed oilseed rape plants with adequate nutrition, glucosinolate biosynthesis and accumulation in leaves follows a consistent pattern. The rate of accumulation is high in young leaves, and the glucosinolate content and concentration are highest as the leaf reaches full expansion

(Porter *et al.*, 1991). There is then a decline in leaf glucosinolate content, indicating metabolism of glucosinolates, or possibly even transport out of the leaf (though this has yet to be demonstrated). Older leaves, particularly as they approach senescence, have a very low glucosinolate content.

However, a variety of stresses and environmental factors can perturb this pattern. We have found that low nutrient levels or low light intensity reduce the glucosinolate content of rape leaves. On the other hand, water stress (drought or flooding) seems to increase the glucosinolate content in *Arabidopsis* plants (Porter, Kiddle & Wallsgrove, unpublished observations).

The response of glucosinolate-containing plants to pest and pathogen attack is well documented, and a marked increase in glucosinolate content in attacked or damaged plants has been noted. Insect attack and mechanical damage (Birch, Griffiths & Smith, 1990; Koritsas, Lewis & Fenwick, 1991; Bodnaryk, 1992), and mammalian herbivory (MacFarlane-Smith, Griffiths & Boag, 1991) have all been found to increase glucosinolate content, particularly indolyl glucosinolates. Fungal infection also increases the content of indolyl glucosinolates in rape leaves, but all other classes of glucosinolate also increase (Doughty *et al.*, 1991). The increase in alkenyl compounds, though large, is transitory, and a similar response may have been missed in other studies where samples were taken at one fixed time after application of the stress or treatment.

The signalling mechanism responsible for such changes, and the underlying biochemistry of the stimulation of glucosinolate synthesis, are unknown. Some abiotic elicitors with known physiological effects in plants cause an increase in glucosinolate content of rape leaves. Methyl jasmonate, applied as a spray, as vapour, or by injection into leaves, increases the content of indolyl glucosinolates markedly (Bodnaryk, 1994; Doughty *et al.*, 1994), though other glucosinolates are unaffected. Salicylic acid, applied as a soil drench, stimulates the accumulation of phenylethylglucosinolate in rape leaves (Kiddle *et al.*, 1994), with a much smaller effect on indolyl compounds and no effect on alkenyl glucosinolates. Ethylene has no apparent effect (Kiddle, unpublished observations).

So, although some putative plant signalling compounds do influence glucosinolate synthesis and accumulation, none of those yet tested fully mimic the response to pest and pathogen attack (particularly fungal infection). Preliminary research in our laboratory has been unable to detect significant increases in glucosinolate biosynthetic enzyme activities during attack- or elicitor-mediated glucosinolate accumulation, so the biochemical mechanism for the enhanced accumulation is unknown. More work in this area is clearly necessary.

Conclusions

We are only at the early stages of understanding the biochemistry of glucosinolate synthesis. Previous assumptions concerning the pathway which relied on analogy to cyanogenic glucoside biosynthesis are almost certainly mistaken. At least with respect to glucosinolate synthesis in *Brassica* species, there are no common features to the two pathways beyond a certain similarity in pathway intermediates – though these are generated by quite different biochemical systems. The biosynthetic pathway(s) in non-Crucifers may well be rather different, with possibly more similarity to the cyanogenic glucoside pathway, though this needs to be investigated much more thoroughly.

Now that cell-free enzyme systems for the early steps in glucosinolate biosynthesis have been developed, progress in characterizing these enzymes should be rapid. In conjunction with the genetic studies described in the next chapter, we may soon be able to describe the pathway in some detail in terms of its biochemistry and molecular genetics, and clear up the remaining confusions concerning the precise chemical nature of the intermediates. Much remains to be done to uncover the regulatory mechanisms operating at all levels, particularly those controlling the increase in glucosinolate accumulation which follows pest and pathogen attack.

References

Benn, M.H. (1962). Biosynthesis of the mustard oils. *Chemistry and Industry (London)*, 1907.

Bennett, R., Donald, A., Dawson, G., Hick, A. & Wallsgrove R.M. (1993). Aldoxime-forming microsomal enzyme systems involved in the biosynthesis of glucosinolates in oilseed rape leaves. *Plant Physiology*, **102**, 1307–12.

Birch, A.N.E., Griffiths, D.W. & Smith, W.H.M. (1990). Changes in forage and oilseed rape glucosinolates in response to attack by turnip root fly (*Delia floralis*). *Journal of Science of Food and Agriculture*, **51**, 309–20.

Bodnaryk, R.P. (1992). Effects of wounding on glucosinolates in the cotyledons of oilseed rape and mustard. *Phytochemistry*, **31**, 2671–77.

Bodnaryk, R.P. (1994). Potent effect of jasmonates on indole glucosinolates in oilseed rape and mustard. *Phytochemistry*, **35**, 301–5.

Chew, F.S. (1988). Biological effects of glucosinolates. In *Biologically Active Natural Products*, ed. H.G.Cutler, pp. 155–81. Washington DC: American Chemical Society.

Chisolm, M.D. & Wetter, L.R. (1966). Biosynthesis of mustard oil glucosides. VII. Formation of sinigrin in horseradish from homomethionine-2-^{14}C and homoserine-2-^{14}C. *Canadian Journal of Biochemistry*, **44**, 1625–32.

Coecke, S., Mertens, K., Segaert, A., Callaerts, A., Vercruysse, A. and Rogiers, V. (1992). Spectrophotometric measurement of flavin-containing monooxygenase activity in freshly isolated rat hepatocytes and their culture. *Analytical Biochemistry*, **205**, 285–8.

Conn, E.E. (1981). Cyanogenic glucosides. In *The Biochemistry of Plants*, ed. E.E.Conn, vol. 7, pp. 479–500. New York: Academic Press.

Dawson, G.W., Hick, A.J., Pickett, J.A., Bennett, R.N., Donald, A. & Wallsgrove R.M. (1993). Synthesis of glucosinolate precursors and investigations into the biosynthesis of phenylalkyl- and methylthioalkylglucosinolates. *Journal of Biological Chemistry*, **268**, 27154–9.

Doughty, K.J., Porter, A.J.R., Morton, A.M., Kiddle, G., Bock, C.H. & Wallsgrove R.M. (1991). Variation in the glucosinolate content of oilseed rape leaves. II. Response to infection by *Alternaria brassicae* (Berk) Sacc. *Annals of Applied Biology*, **118**, 469–77.

Doughty, K.J., Kiddle, G.A., Pye, B.J., Wallsgrove, R.M. & Pickett, J.A. (1994). Methyl jasmonate-induced accumulation of glucosinolates in oilseed rape leaves. *Phytochemistry* (in press).

Ettlinger, M.G. & Kjaer, A. (1968). Sulfur compounds in plants. In *Recent Advances in Phytochemistry*, eds. T.J.Mabry, R.E.Alston & V.C.Runeckles, pp. 59–144. New York: Appleton-Century-Crofts.

Fenwick, G.R., Heaney, R.K. & Mullin, J. (1983). Glucosinolates and their breakdown products in food and food plants. *CRC Critical Reviews in Food Science and Nutrition*, **18**, 123–201.

Fieldsend, J. & Milford, G.F.J. (1994). Changes in glucosinolates during crop development in single- and double-low genotypes of winter oilseed rape: I. Profiles and tissue water concentrations in vegetative tissues and developing pods. *Annals of Applied Biology*, (in press).

Giovanelli, J.G., Mudd, S.H. & Dakto, A. (1980). Sulphur amino acids in plants. In *The Biochemistry of Plants*, eds. P.K.Stumpf & E.E.Conn, vol. 5, pp. 453–505. San Diego: Academic Press.

Glendening, T.M. & Poulton, J.E. (1988). Glucosinolate biosynthesis. *Plant Physiology*, **86**, 319–21.

Glendening, T.M. & Poulton, J.E. (1990). Partial purification and characterisation of a 3'-phosphoadenosine 5'-phosphosulphate: desulphoglucosinolate sulphotransferase from cress. *Plant Physiology*, **94**, 811–18.

Grootwassink, J.W.D., Balsevich, J.J. & Kolenovsky, A.D. (1990). Formation of sulfatoglucosides from exogenous aldoximes in plant cell cultures and organs. *Plant Science,* **66,** 11–20.

Halkier, B.A., Olsen, C.E. and Møller, B.L. (1989). The biosynthesis of cyanogenic glucosides in higher plants. *Journal of Biological Chemistry,* **33,** 19487–94.

Hall, D.I., & Smith, I.K. (1983). Partial purification and characterisation of cystine lyase from cabbage (*Brassica oleracea var.* Capital). *Plant Physiology,* **72,** 654–8.

Hamamato, A., & Mazelis, M. (1986). The C-S lyases of higher plants: isolation and properties of homogeneous cystine lyase from brocolli (*Brassica oleracea var.* Botrytis) buds. *Plant Physiology,* **80,** 702–6.

Helmlinger, J., Rausch, T. & Hilgenberg, W. (1987). A soluble protein factor from Chinese cabbage converts indole-3-acetaldoxime to IAA. *Phytochemistry,* **26,** 615–8.

Jain, J.C., GrootWassinck, J.W.D., Reed, D.W. & Underhill, E.W. (1990). Persistent co-purification of enzymes catalysing the sequential glucosylation and sulfation steps in glucosinolate biosynthesis. *Journal of Plant Physiology,* **136,** 356–61.

Kiddle, G.A., Doughty, K.J. & Wallsgrove, R.M. (1994). Salicylic acid induced accumulation of glucosinolates in oilseed rape leaves. *Journal of Experimental Biology* (in press).

Kindl, H. (1964). Zür biosynthese des sinalbins. *Monat schefte für Chemie,* **95,** 439–48.

Kindl, H. (1965). The biosynthesis of sinalbin. II. Synthesis of sinalbin in mustard plants of various ages. *Monatschefte für Chemie,* **96,** 527–32.

Kindl, H. & Underhill, E.W. (1968). Biosynthesis of mustard oil glucosides: N-hydroxyphenylalanine, a precursor of glucotropaeolin and a substrate for the enzymatic and non-enzymatic formation of phenylacetaldehyde aldoxime. *Phytochemistry,* **7,** 745–56.

Kjaer, A. & Larsen, P.O. (1973). Non-protein amino-acids, cyanogenic glycosides, and glucosinolates. In *Biosynthesis,* ed. T.A. Giessman, vol. 2, pp. 71–105. London: The Chemical Society.

Kjaer, A. & Larsen, P.O. (1976). Non-protein amino acids, cyanogenic glucosides, and glucosinolates. In *Biosynthesis,* ed. T.A. Giessman, vol. 4, pp. 179–203. The Chemical Society, London.

Kjaer, A. & Larsen, P.O. (1977). Non-protein amino acids, cyanogenic glucosides, and glucosinolates. In *Biosynthesis,* ed. T.A. Giessman, vol. 5, pp. 120–35. London: The Chemical Society.

Kjaer, A. & Larsen, P.O. (1980). Non-protein amino acids, cyanogenic glucosides, and glucosinolates. In *Biosynthesis,* ed. T.A. Giessman, vol. 6, pp. 155–80. London: The Chemical Society.

Koritsas, V.M., Lewis, J.A. & Fenwick, G.R. (1991). Glucosinolate responses of oilseed rape, mustard and kale to mechanical wounding and infestation by cabbage stem flea beetle. *Annals of Applied Biology*, **118**, 209–21.

Kutachek, M., Prochazka, Z., & Veres, K. (1962). Biogenesis of glucobrassicin, the *in vitro* precursor of ascorbigen. *Nature, London*, **104**, 393–4.

Larsen, P.O. (1981). Glucosinolates. In *The Biochemistry of Plants*, ed. E.E.Conn, vol. 7, pp. 501–25.

Ludwig-Muller, J. & Hilgenberg, W. (1988). A plasma membrane-bound enzyme oxidises L-tryptophan to indole-3-acetaldoxime. *Physiologia Plantarum*, **74**, 240–50.

Ludwig-Muller, J., Rausch, T., Lang, S. & Hilgenberg, W. (1990). Plasma membrane-bound high pI isoenzymes convert tryptophan to indole-3-acetaldoxime. *Phytochemistry*, **29**, 1397–400.

Lykkesfeldt, J. & Møller, B.L. (1993). Synthesis of benzylglucosinolate in *Tropaeolum majus* L. isothiocyanates as potent enzyme inhibitors. *Plant Physiology*, **102**, 609–13.

MacFarlane-Smith, W.H., Griffiths, D.W. & Boag, B. (1991). Overwintering variation in glucosinolate content of green tissues of rape *(Brassica napus)* in response to grazing by wild rabbit *(Oryctolagus cuniculus)*. *Journal of the Science of Food and Agriculture*, **56**, 511–21.

Mannervik, B., and Danielson, U.H. (1988). Glutathione transferases – structure and catalytic activity. *CRC Critical Reviews in Biochemistry*, **23**, 283–337.

Matsuo, M., Kirkland, D.F. & Underhill, E.W. (1972). 1-Nitro-2-phenylethane, a possible intermediate in the biosynthesis of benzylglucosinolate. *Phytochemistry*, **11**, 697–701.

Mazelis, M., Scott, K. & Gallie, D. (1982). Non-identity of cystine lyase with cystathionase in turnip roots. *Phytochemistry*, **5**, 991–5.

Pickett, C.B. & Lu, A.Y.H. (1989). Glutathione-S-transferases: gene structure, regulation and biological function. *Annual Review of Biochemistry*, **58**, 743–64.

Porter, A.J.R., Morton, A.M., Kiddle, G., Doughty, K.J. & Wallsgrove, R.M. (1991). Variations in the glucosinolate content of oilseed rape leaves. I. Effect of leaf age and position. *Annals of Applied Biology*, **118**, 461–7.

Poulton J.E. & Møller, B.L. (1993). Glucosinolates. In *Methods in Plant Biochemistry*, Vol.9, ed. P.J.Lea, pp. 209–37. London: Academic Press.

Rodman JE. (1981). In *Phytochemistry and Angiosperm Phylogeny*, ed. D.A.Young & D.S.Seigler, pp 43–79. New York: Praeger.

Rossiter, J.T., James, D.C. & Atkins, N. (1990). Biosynthesis of 2-hydroxy-3-butenylglucosinolates and 3-butenylglucosinolate in *Brassica napus*. *Phytochemistry,*, **29**, 2509–12.

Schraudolf, H. and Bergmann, F. (1965). Metabolism of indole derivatives in *Sinapis alba* L. *Planta*, **67**, 75–95.

Sorensen, H. (1991). Glucosinolates: structure – properties – function. In *Canola and Rapeseed*, ed. F.Shahidi, pp. 149–72. New York: Van Nostrand Reinhold.

Staton, A.L. & Mazelis, M. (1991). The C–S lyases of higher plants: homogeneous cystathionase of spinach leaves. *Archives of Biochemistry and Biophysics*, **290**, 46–50.

Tapper, B.A. & Butler, G.W. (1967). Conversion of oximes to mustard oil glucosides (glucosinolates). *Archives of Biochemistry and Biophysics,,* **120**, 719–21.

Timmerman, K.P. (1989). Molecular characterisation of corn glutathione S-transferase isozymes involved in herbicide detoxification. *Physiologia Plantarum*, **77**, 465–71.

Umbarger, H.E. (1978). Amino acid biosynthesis and its regulation. *Annual Review of Biochemistry*, **47**, 533–606.

Underhill, E.W. (1965). Biosynthesis of mustard oil glucosides. *Canadian Journal of Biochemistry*, **43**, 179–87.

Underhill, E.W. (1967). Biosynthesis of mustard oil glucosides. *European Journal of Biochemistry*, **2**, 61–3.

Underhill, E.W. (1968). Biosynthesis of mustard oil glucosides. *Canadian Journal of Biochemistry*, **46**, 401–5.

Underhill, E.W., Chisolm, M.D. & Wetter, L.R. (1962). Biosynthesis of mustard oil glucosides. *Canadian Journal of Biochemistry*, **40**, 1505–14.

Underhill, E.W. & Chisolm, M.D. (1964). Biosynthesis of mustard oil glucosides. *Biochemical and Biophysical Research Communications*, **14**, 425–30.

Underhill, E.W. and Wetter, L.R. (1969). Biosynthesis of mustard oil glucosides. *Plant Physiology*, **44**, 584–590.

Underhill, E.W., Wetter, L.R. & Chisholm, M.D. (1973). Biosynthesis of glucosinolates. *Biochemical Society Symposia*, **38**, 303–26.

Wallsgrove, R.M., Lea, P.J. & Miflin, B.J. (1983). Intracellular localisation of aspartate kinase and the enzymes of threonine and methionine biosynthesis in green leaves. *Plant Physiology*, **71**, 780–4.

Wetter, L.R. & Chisolm, M.D. (1968). Sources of sulfur in the thioglucosides of various higher plants. *Canadian Journal of Biochemistry*, **46**, 931–935.

Ziegler, D.M. (1988). Flavin-containing monooxygenases: catalytic mechanism and substrate specificity. *Drug Metabolism Reviews*, **19**, 1–32.

RICHARD MITHEN and DIKRAN TOROSER

Biochemical genetics of aliphatic glucosinolates in *Brassica and Arabidopsis*

Aliphatic glucosinolates are sulphonated thioglucosides which are found in all hitherto investigated plants of the order Capparales, which includes the genera *Brassica* and *Arabidopsis*. Although these secondary metabolites share the general structure (see previous chapter and Fig. 1), differences in their 'R' substituents enable classification into several distinct structural classes. The aliphatic glucosinolates found in *Brassica* typically include methylthioalkyl, methylsulphinylalkyl, alkenyl and hydroxyalkenyl homologues of propyl, butyl and pentyl glucosinolates. *A. thaliana* contains a similar variety of glucosinolates, with the addition of hydroxyalkyl glucosinolates and trace levels of longer side-chain homologues (Hogge *et al.*, 1988). Following disruption of cell integrity, aliphatic glucosinolates are hydrolysed by endogenous β-thioglucoside glucohydrolases ('myrosinases'; EC 3.2.3.1) to give a complex mixture of products, of which D-glucose, sulphate and isothiocyanates are major components (Cole, 1976; Benn, 1977; Fenwick, Heaney & Mullin, 1983). In addition to being responsible for the characteristic flavour of cruciferous condiments, hydrolytic products of glucosinolates such as isothiocyanates, thiocyanates and nitriles are known to impart antinutritional properties to protein-rich oilseed 'meals' commonly used as animal-feed (Bell, Benjamin & Giovanetti, 1972; Fenwick, 1984). In particular, isothiocyanate derivatives of hydroxyalkenyl glucosinolates undergo spontaneous cyclization to produce substituted oxazolidine-2-thiones (Macleod & Rossiter, 1987) which have potent goitrogenic properties (Astwood, Greer & Ettlinger, 1949; Langer, 1966). The occurrence of these antinutritional compounds in seed-meals of major oilseed *Brassica* crops has led to efforts to reduce the level of aliphatic glucosinolates in the seeds of oilseed rape. Following the discovery of the spring rapeseed cultivar 'Bronowski', and its incorporation into conventional breeding programmes, an eight- to ten-fold reduction in the aliphatic glucosinolate levels of oilseed rape was achieved by Canadian plant breeders and led to the development of the so-called '00-

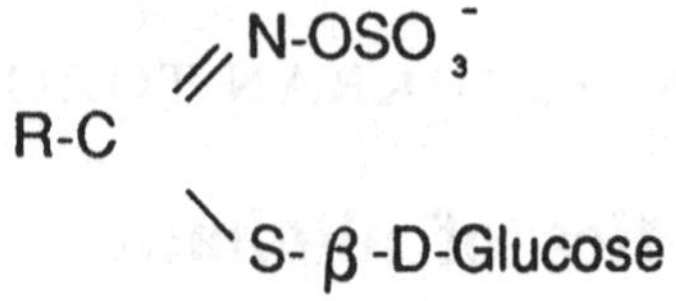

R:

CH_3-S-$(CH_2)_n$-	Methylthioalkyl
CH_3-SO-$(CH_2)_n$-	Methylsulphinylalkyl
CH_3-SO_2-$(CH_2)_n$-	Methylsulphonylalkyl
CH_2=CH-$(CH_2)_n$-	Alkenyl
CH_2=CH-CHOH-$(CH_2)_n$-	Hydroxyalkenyl
HO-$(CH_2)_n$-	Hydroxyalkyl

Fig. 1. Molecular structure of aliphatic glucosinolates.

cultivars', i.e. cultivars with low levels of both erucic acid and glucosinolates in their seed.

Aliphatic glucosinolates and their hydrolytic byproducts have been shown to be important in the interaction of pests and pathogens with cruciferous host species (Greenhalgh & Mitchell, 1976; Blau *et al.*, 1978; Buchwaldt, Nielson & Sorensen, 1985). Specialized pests of the Brassicaceae are attracted to plants containing glucosinolates and are thought to use these metabolites and their hydrolytic products as feeding and oviposition stimuli (Nair & McEwen, 1976; Chew, 1988; Bartlett & Williams, 1989; Reed, Pivnick & Underhill *et al.*, 1989). In contrast, non-specialized pests such as slugs and birds are repelled by plants which contain high levels of alkenyl glucosinolates (Glen, Jones & Fieldsend, 1990; Mithen, 1992). These effects may be attributed to the pungency and cytotoxicity of volatile byproducts which arise from hydrolysis of glucosinolates by myrosinases (Horakova, 1966).

Biosynthesis of aliphatic glucosinolates

Several reviews of the biosynthesis of aliphatic glucosinolates have been published (Underhill, Wetter & Chisholm, 1973; Underhill, 1980; Fenwick *et al.,* 1983; Poulton & Moller, 1993). Their biosynthetic derivation from amino acids was demonstrated in a number of crucifers with the use of ^{14}C, ^{15}N and ^{35}S-labelled precursor studies (Kjaer & Larsen, 1973; 1976; Larsen, 1981 and references therein). A model for the biosynthesis of aliphatic glucosinolates based on biochemical and genetic evidence is shown in Fig 2. Further details of side chain elongation and modification are presented in Figs. 3 and 4. The biosynthetic model shown in Fig. 2 can be considered as being composed of three integral parts. First, the synthesis of homologues of methionine, which can act as intermediates for the biosynthesis of glucosinolates with different side chain substituents. Secondly, the synthesis of the glycone moiety of glucosinolates (discussed in detail in the preceding chapter), and, thirdly, a number of side chain modifications at the glucosinolate level. While the scheme shown in Fig. 2 provides a useful working model, several parts of the metabolic pathway await clarification at the biochemical level.

Only a few studies have focused on the genetic regulation of aliphatic glucosinolate biosynthesis. To date, most reports have concentrated on the factors which determine the low levels of seed aliphatic glucosinolates in various oilseed rape cultivars. Limited attention has been paid to the genetic basis of the biosynthesis of individual aliphatic glucosinolate classes occurring in *Brassica,* and related crucifers. A detailed understanding of this genetic regulation is a prerequisite in attempts to modify the glucosinolate profiles of major *Brassica* crops. In this context, the elimination of glucosinolates responsible for the attraction of specialized pests and the retention (and addition to) of those which assist in protection against non-specialized pests must be a long-term goal of research efforts.

Amongst the reasons for the lack of information on the genetic basis of aliphatic glucosinolate biosynthesis is the dearth of natural variation in the aliphatic glucosinolate profile of *B.napus*. In this species, the total levels of aliphatic glucosinolates may vary, but the ratio of individual aliphatic glucosinolates are similar within specific tissues (Fig. 5). In contrast to the lack of variation in *B.napus,* considerable variation can be found in wild and cultivated forms of its diploid progenitors.

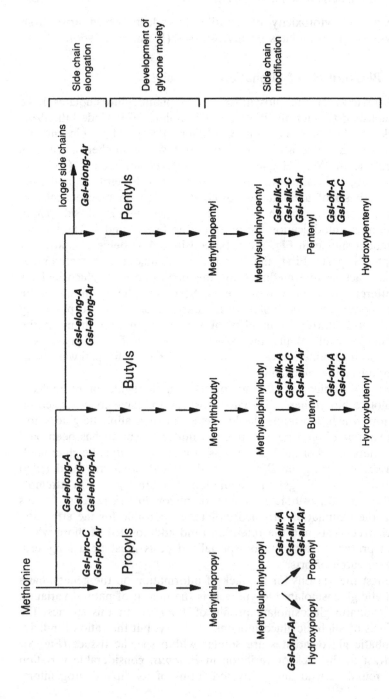

Fig. 2. A model of aliphatic glucosinolate biosynthesis. Details of the gene symbols are described in the text. The suffixes -A, -C and -Ar refer to the *Brassica* A, *Brassica* C and *A.thaliana* genomes, respectively.

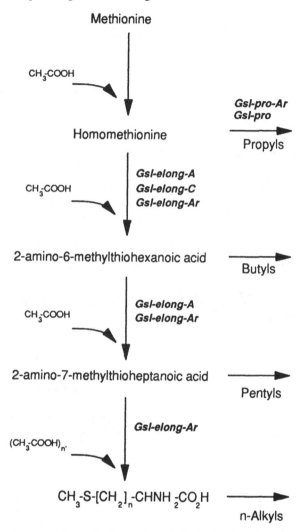

Fig. 3. Side chain elongation of aliphatic glucosinolate precursors. The elongation process is thought to occur by the addition of the methyl carbon of acetate to methionine and side chain elongated homologues. Allelic variation at the *Gsl-elong* loci regulate the extent of elongation.

This variation has been used to analyse the genetic basis of aliphatic glucosinolate biosynthesis in *Brassica* species, either by crosses between diploid accessions or by the development of synthetic amphidiploid lines of *B.napus* prior to crosses to natural forms of *B.napus* (Gland,

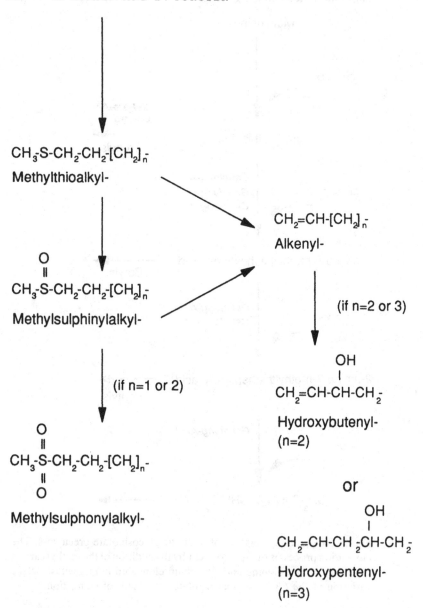

Fig. 4. Conversion of methylthioalkyl and/or methylsulphinylalkyl glucosinolates to alkenyl glucosinolates. The precise pathway *in planta* requires further investigation.

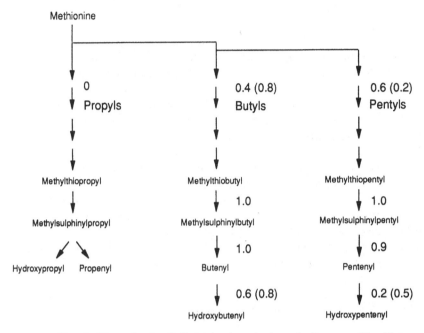

Fig. 5. Biosynthesis of aliphatic glucosinolates in *B.napus*. The Figure represent the proportion of metabolites which flow through the different parts of the pathway in leaves of winter oilseed rape, while those in parentheses represent the proportion in seeds. In leaves, pentenyl glucosinolates predominate, while in seeds hydroxybutenyl glucosinolates are dominant. Substrate flow through the pathway is regulated by the alleles at the *Gsl* loci (Fig. 2).

1981; Chen & Heneen, 1989). Considerable diversity also exists in the aliphatic glucosinolate profiles of ecotypes of *Arabidopsis thaliana* (Mithen, unpublished data), in addition to mutation induced variation (Haughn *et al.*, 1991). Genetic and biochemical evidence from both genera is useful for the development of general models of the genetic regulation of glucosinolate biosynthesis. In the present chapter, three aspects of the genetic regulation of glucosinolate biosynthesis are considered. First the regulation of side chain elongation, secondly the regulation of side chain modifications and thirdly the genetic basis of the seed-glucosinolate content of *B.napus*. Finally, novel approaches to the genetic modification of aliphatic glucosinolates in *Brassica* crops are discussed.

Side chain elongation

Biochemistry

In the biosynthesis of glucosinolates, side chain elongation occurs prior to the development of the glycone moiety. Administration of ^{14}C-labelled precursors to numerous crucifers has shown that this may be achieved by the successive addition of the methyl carbon of acetate to methionine, and its side chain elongated homologues. As a result, a series of sulphur-containing amino acids are thus produced (R = $CH_3S(CH_2)_n$; where $n = 3$ to 8) which can act as precursors in the development of the aglycone moieties of aliphatic glucosinolates (Chisholm & Wetter, 1966; Underhill *et al.*, 1973; Fig. 3). It is possible that a multienzyme reaction system is involved in the transamination of sulphur-containing amino acid precursors to produce corresponding α-keto acids, followed by carbon chain elongation by condensation with acetyl-CoA, and then conversion back to the amino acid homologue, analogous to the biosynthesis of leucine from valine (Chisholm & Wetter, 1967; Haughn *et al.*, 1991).

Variation in side chain length

Accessions of *B.rapa* and *B.oleracea sensu lato (i.e.* including all species of the $n = 9$ C genome complex) exhibit inter- and intra-specific variation in the side chain length of their aliphatic glucosinolates. For example, taxa belonging to the $n = 9$ species complex may contain only propyl glucosinolates (e.g. *B.insularis, B.rupestris, B.macrocarpa),* only butyl glucosinolates *(B.montana, B.insularis)* or both classes of propyl and butyl side chain lengths (cultivated forms of *B.oleracea).* No C genome accession has ever been found to contain significant amounts of pentyl glucosinolates. In contrast, all hitherto analysed accessions of *B.rapa* invariably contain butyl glucosinolates and may or may not contain pentyl glucosinolates (Mithen *et al.*, 1987; Giamoustaris & Mithen, unpublished data). In ecotypes of *A.thaliana,* either propyl glucosinolates may predominate or both propyl and butyl glucosinolates may be present. Trace levels of longer chain homologues (up to $n = 8$) can also be detected in many ecotypes.

Inheritance of side chain length

The inheritance of side chain length has been studied in crosses between natural forms of *B.napus* and synthetic *B.napus* lines with contrasting glucosinolate profiles (Magrath *et al.*, 1993, 1994). Alleles at a single locus in the C genome *(Gsl–pro)* determine the presence or absence of propyl glucosinolates, while alleles at two other loci *(Gsl–elong–A* and *Gsl–elong–C;* Fig. 2) determine the extent of side chain elongation. Alleles at the *Gsl–elong–C* locus may result either in an absence of side chain elongated glucosinolates or produce butyl glucosinolates, while alleles at the *Gsl–elong–A* locus produce either butyl glucosinolates or both butyl and pentyl glucosinolates. Further studies with crosses between accessions of *B.oleracea* which vary in their side chain length of aliphatic glucosinolates provided further support that side chain elongation is regulated by a single locus in diploid species and two loci in amphidiploid species (Giamoustaris and Mithen, unpublished observations). Similarly, crosses between *A.thaliana* ecotypes suggest that alleles at a single locus *(Gsl–elong–Ar)* regulate the production of butyl glucosinolates as opposed to propyl glucosinolates (Magrath *et al.*, 1994).

The extent of side chain elongation may depend upon the substrate specificity of gene products of particular alleles at the *Gsl–elong* loci. If the gene products can regulate the addition of carbon to both homomethionine and to 2-amino-6-hexanoic acid, the precursors of butyl and pentyl glucosinolates may be produced, whereas if the pertinent gene products cannot recognise any side chain elongated homologues of methionine only propyl glucosinolates will be produced (Fig. 3). The precise biochemical function and substrate specificity of the *Gsl–elong* loci gene products has not been investigated. An experimental approach for elucidation of the nature of the gene products of the *Gsl–elong* loci may include cloning of this gene from *A.thaliana* through a map-based cloning strategy coupled with structural characterization of the isolated gene products.

In the current model of glucosinolate biosynthesis, the precursors of propyl glucosinolates (homomethionine or the corresponding α-keto acid) are also precursors of longer chain aliphatic glucosinolates. If this model is correct, it is likely that the gene product of the *Gsl–pro* locus is not involved in side chain elongation but in the subsequent development of the glycone moiety (Fig.3).

Side chain modifications

Methylthioalkyl, methylsulphinylalkyl and alkenyl glucosinolates

Aliphatic glucosinolates of different side chain lengths may have similar types of side chain modifications. The initial products of biosynthesis are probably methylthioalkyl glucosinolates. Subsequent oxidation of the thiol will produce methylsulphinylalkyl and methylsulphonylalkyl homologues, while removal of the methylthio group (or a group with the sulphur at a higher level of oxidation), followed by desaturation, will lead to alkenyl homologues (Fig. 4). Radiotracer studies have demonstrated the reversible formation of methylthiopropyl and methyl-sulphinylpropyl glucosinolates in *Cheiranthus kewensis* (Chisholm, 1972). It has also been shown that both compounds can act as precursors of propenyl glucosinolates (Chisholm & Matsuo, 1972). Which compound is the immediate precursor of propenyl glucosinolates *in planta* remains to be elucidated. It is possible that alkenyl glucosinolates may be derived from both methylthioalkyl and methylsulphinylalkyl glucosinolates.

As with side chain length, there is no variation in *B.napus* for the types of side chain modifications. All genotypes of this species have a mixture of alkenyl and hydroxyalkenyl glucosinolates, with trace levels of methylsulphinylpentyl glucosinolate (Fig. 5). In contrast, cultivated forms of *B.oleracea* contain high levels of methylsulphinylpropyl and methylsulphinylbutyl glucosinolates. Wild forms of this species may either have predominantly alkenyl glucosinolates (e.g. *B.montana* or *B.cretica*), or methylthioalkyl and methylsulphinylalkyl glucosinolates (*B.villosa, B.macrocarpa* and other taxa from Southern Italy). *A.thaliana* ecotypes exhibit variation for these classes of glucosinolates. For example, methylsulphinylbutyl glucosinolate predominates in Colombia, whereas Stockholm has significant amounts of butenyl glucosinolate.

Genetic studies of diploid Brassicas suggest that alleles at a single locus *(Gsl–alk)* regulate the conversion of methylsulphinylalkyl (or methylthioalkyl) glucosinolates into alkenyl glucosinolates. In *B.napus*, there are two copies of this genes, one from the A and one in the C genome (Magrath *et al.*, 1993). The products of this single locus are able to convert methylthioalkyl and/or methylsulphinylalkyl homologues of each of the three side chain lengths into alkenyl glucosinolates, although with different efficiencies (Magrath *et al.*, 1993). Similarly, in *A.thaliana* a single gene *(Gsl–alk–Ar)* converts methylsulphinylpropyl and methylsulphinylbutyl glucosinolates into propenyl and butenyl glucosinolates respectively (unpublished data).

Hydroxyalkenyl and hydroxyalkyl glucosinolates

Studies with radioisotopic precursors suggest that the hydroxylation of butenyl glucosinolates occurs as the last step in their biosynthesis (Rossiter, James & Atkins, 1990). This observation is supported by genetic studies of crosses between genotypes which differ in their extent of hydroxylation (Magrath *et al.*, 1993). As with other parts of the glucosinolate pathway, it has been shown that in diploid *Brassica* species alleles at a single locus *(Gsl–oh)* regulate the hydroxylation of either butenyl or pentenyl glucosinolates (but not propenyl glucosinolates). In *B.napus,* there are two copies of this locus, one from the A and one from the C genome. All natural forms of *B.napus* have functional alleles at the two *Gsl–oh* loci, but null alleles can be introduced from diploid *Brassica* species to prevent the production of hydroxyalkenyl glucosinolates (Magrath *et al.*, 1993; Parkin *et al.*, 1994). The level of hydroxyalkenyl glucosinolates observed in the seeds is correlated with the extent of hydroxylation in vegetative tissues, although the alleles at the *Gsl–oh–C* locus have a much greater effect than the alleles at the *Gsl–oh–A* locus (Parkin *et al.*, 1994).

Hydroxyalkenyl glucosinolates also occur in *A.thaliana,* but hydroxy-propyl and hydroxybutyl glucosinolates are more abundant. In the ecotype Landsberg *erecta,* hydroxypropyl glucosinolates comprise over 95% of the total aliphatic glucosinolates. In crosses with Colombia, which has predominantly methylsulphinylbutyl glucosinolates, there is segregation at two loci, one *(Gsl–elong–Ar)* which determines the side chain length (see above), and a second *(Gsl–ohp–Ar)* which regulates the conversion of methylsulphinylpropyl to hydroxypropyl (but does not result in the conversion of methylsulphinylbutyl to hydroxybutyl).

Genetic regulation of glucosinolates in seeds of oilseed rape

One of the important factors contributing to the success of oilseed rape *(B.napus)* as a major crop in the developed world has been the success of plant breeders in producing cultivars with low levels of erucic acid and glucosinolates. The first attempts to produce low-glucosinolate cultivars were unsuccessful. While reductions in seed glucosinolate levels was achieved, the lines were agronomically inadequate, showing unacceptable susceptibility to grazing by birds and mammals. Subsequent studies have shown that this was due to the reduction of aliphatic glucosinolates in all parts of the plant which resulted in the vegetative tissue becoming highly palatable to pests (Mithen, 1992). In currently grown low-glucosinolate cultivars, the reduction in glucosinolate levels has been restricted to the propagative tissues (Milford *et al.*, 1989; Mithen, 1992).

The physiological basis of the differences in the level of seed glucosinolates found in high- and low-glucosinolate cultivars remains unexplained. To date most reports suggest that aliphatic glucosinolates are not synthesized by the developing embryo but are transported into the developing seed from the pod wall, probably as intact glucosinolates (Magrath & Mithen, 1993). It seems likely that there is a block in the biosynthesis of glucosinolates specifically within the pod wall so that there are smaller amounts of glucosinolate available for transport into the developing seed (Josefsson, 1973). This character is under complex genetic control. Levels of glucosinolates segregate in a continuous manner in F_2 and backcross populations suggesting the action of several genes (Rucker & Rudolf, 1992).

Genetic modification of glucosinolate content

Future attempts to modify the glucosinolate profiles of *Brassica* crops may have two objectives. First, to further reduce the total level of aliphatic glucosinolates within seeds, and, secondly, to seek specific changes in the types of individual glucosinolates within both leaves and seeds. Exotic breeding material such as synthetic *B.napus* lines may be used in crosses to introduce novel alleles at the relevant *Gsl* loci. For example, development of low progoitrin cultivars of *B.napus* can be achieved by introgressing alleles at the two *Gsl–oh* loci (Parkin *et al.*, 1994). Marker-assisted backcrossing programmes with either RFLP or RAPD markers for the novel *Gsl* alleles will increase the rate at which they can be introduced into oilseed rape cultivars.

As an alternative strategy, it should be possible to clone genes from either *Brassica* or *Arabidopsis* which regulate specific steps in aliphatic glucosinolate biosynthesis. As products of many of these genes are either not known or difficult to purify, a map-based cloning strategy from *Arabidopsis* may be the most efficient method of isolating these genes. The *Arabidopsis* clones may then be used as probes to identify *Brassica* homologues or used directly in transformation experiments with oilseed rape cultivars. The level of total aliphatic glucosinolates may be reduced by transformation with an antisense construct of one of several genes which are important in the biosynthetic pathway, such as those in side chain elongation. In order to achieve a reduction in seed glucosinolate levels without affecting the level of leaf glucosinolates which are important in defence, the antisense construct will need to be driven by a pod specific promoter, the pod being the site of biosynthesis of seed aliphatic glucosinolates. It may also be possible to use antisense constructs of genes which regulate specific steps in the biosynthetic pathway, such as hydroxylation of alkenyl glucosinolates and the removal of the terminal methylthio group, to modify the types of individual glucosinolates which occur in oilseed rape.

References

Astwood, E.B., Greer, M.A. & Ettlinger, M.G. (1949). 1-5-Vinyl-2-thiooxazolidone, an antithyroid compound from yellow turnip and from *Brassica* seeds. *Journal of Biological Chemistry,* **181,** 121–30.

Bartlett, E. & Williams, I.H. (1989). Host plant selection by the cabbage stem flea beetle *(Psylliodes chrysocephala).* In *Production and Protection of Oilseed Rape and Other Brassica Crops. Aspects of Applied Biology,* **23,** 335–8.

Bell, J.M., Benjamin, B.R. & Giovanetti, P.M. (1972). Histopathology of thyroids and livers of rats and mice fed diets containing *Brassica* glucosinolates. *Canadian Journal of Animal Science,* **52,** 395–406.

Benn, M. (1977). Glucosinolates. *Pure and Applied Chemistry,* **49,** 197–210.

Blau, P.A., Feeny, P., Contardo, L. & Robson, D.S. (1978). Allylglucosinolate and herbivorous caterpillars: a contrast in toxicity and tolerance. *Science,* **200,** 1296–8.

Buchwaldt, L., Nielsen, J. K. & Sorensen, H. (1985). Preliminary investigations of the effect of sinigrin on *in vitro* growth of three fungal pathogens of oilseed rape. In *Advances in the Production and Utilization of Cruciferous Crops,* ed. H. Sorensen, pp. 260–7. Martinus Nijhoff Publishers.

Chen, B.-Y. & Heneen, W.K. (1989). Resynthesized *Brassica napus L.:* a review of its potential in breeding and genetic analysis. *Hereditas,* **111,** 255–63.

Chew, F.S. (1988). Biological effects of glucosinolates. In *Biologically Active Natural Products Potential Use in Agriculture,* ed. H.G. Cutler, pp. 155–81, Washington DC: American Chemical Society Symposium.

Chisholm, M.D. (1972). Biosynthesis of 3-methylthiopropylglucosinolate and 3-methylsulfinylpropylglucosinolate in wallflower *Cheiranthus kewensis. Phytochemistry,* **11,** 197–202.

Chisholm, M.D. & Matsuo, M. (1972). Biosynthesis of allylglucosinolate and 3-methylthiopropylglucosinolates in horseradish, *Armoracia lapathifolia. Phytochemistry,* **11,** 203–7.

Chisholm, M. D. & Wetter, L. R. (1966). Biosynthesis of mustard oil glucosides. VII. Formation of sinigrin in horseradish from homomethionine-2-^{14}C and Homoserine-2-^{14}C. *Canadian Journal of Biochemistry,* **44,** 1625–32.

Chisholm, M. D. & Wetter, L. R. (1967). The biosynthesis of some isothiocyanates and oxazolidinethiones in rape *(Brassica campestris L.). Plant Physiology,* **42,** 1726–30.

Cole, R.A. (1976). Isothiocyanates, nitriles and thiocyanates as prod-

274   R. MITHEN AND D. TOROSER

ucts of autolysis of glucosinolates in Cruciferae. *Phytochemistry*, **15**, 759–62.

Fenwick, G. R. (1984). Rapeseed as an animal feeding stuff – the problems and analysis of glucosinolates. *Journal of the Association of Public Analysts*, **22** (4), 117–30.

Fenwick, G.R., Heaney, R.K. & Mullin, W.J. (1983). Glucosinolates and their breakdown products in food and food plants. *Critical Reviews in Food Science and Nutrition*, **18**, 123–301.

Gland, A. (1981). Content and pattern of glucosinolates in resynthesised rapeseed. In *Production and Utilization of protein in Oilseed Crops*, ed. E.S. Bunting, pp. 127–35. Martinus Nijhoff Publishers.

Glen, D.M., Jones, H. & Fieldsend, J.K. (1990). Damage to oilseed rape seedlings by the field slug *Deroceras reticulatum* in relation to glucosinolate concentration. *Annals of Applied Biology*, **117**, 197–207.

Greenhalgh, J.R. & Mitchell, N.D. (1976). The involvement of flavour volatiles in the resistance to downy mildew of wild and cultivated forms of *Brassica oleracea*. *New Phytologist*, **77**, 391–8.

Haughn, G.W., Davin, L., Giblin, M. & Underhill, E.W. (1991). Biochemical genetics of plant secondary metabolites in *Arabidopsis thaliana*. The glucosinolates. *Plant Physiology*, **97**, 217–26.

Hogge, L.R., Reed, D.W., Underhill, E.W. & Haughn, G.W. (1988). HPLC separation of glucosinolates from leaves and seeds of *Arabidopsis thaliana* and their identification using thermospray liquid chromatography-mass spectrometry. *Journal of Chromatographic Science*, **26**, 551–6.

Horakova, K. (1966). Cytotoxicity of natural and synthetic isothiocyanates. *Naturwissenschaften*, **15**, 383–4.

Josefsson, E. (1973). Studies of the biochemical background to differences in glucosinolate content in *Brassica napus* L. III. Further studies to localize metabolic blocks. *Physiologia Plantarum*, **29**, 28–32.

Kjaer, A. & Larsen, P.O. (1973). Non-protein amino-acids, cyanogenic glycosides, and glucosinolates. In *Biosynthesis*, ed. T.A. Giessman, vol. 2, pp. 71–105. London: The Chemical Society.

Kjaer, A. & Larsen, P.O. (1976). Non-protein amino-acids, cyanogenic glycosides, and glucosinolates. In *Biosynthesis*, ed. T.A. Giessman, vol. 4, pp. 179–203. London: The Chemical Society.

Langer, P. (1966). Antithyroid action in rats of small doses of some naturally occurring compounds. *Endocrinology*, **79**, 1117–22.

Larsen, P.O. (1981). Glucosinolates. In *The Biochemistry of Plants*, eds. P.K.Stumpf & E.E.Conn, vol. 7, pp. 501–25. New York: Academic Press.

Macleod, A.J. & Rossiter, J.T. (1987). Degradation of 2-hydroxybut-3-enylglucosinolate (progoitrin). *Phytochemistry*, **26**, (3), 669–73.

Magrath, R., Herron, C., Giamoustaris, A. & Mithen, R. (1993). The inheritance of aliphatic glucosinolates in *Brassica napus. Plant Breeding,* **111**, 55–72.

Magrath, R., Bano, F., Morgner, M., Parkin, I., Sharpe, A., Lister, C., Dean, C., Turner, J., Lydiate, D. & Mithen, R. (1994). Genetics of aliphatic glucosinolates I. Side chain elongation in *Brassica napus* and *Arabidopsis thaliana. Heredity,* **72**, 290–99.

Magrath, R. & Mithen, R. (1993). Maternal effects on the expression of individual aliphatic glucosinolates in seeds and seedlings of *Brassica napus. Plant Breeding,* **111**, 249–52.

Milford, G.F.J., Fieldsend, J.K., Porter, A.J.R. & Rawlinson, C.J. (1989). Changes in glucosinolate concentrations during the vegetative growth of single- and double-low cultivars of winter oilseed rape. In *Production and Protection of Oilseed Rape and Other Brassica Crops. Aspects of Applied Biology,* **23**, 83–90.

Mithen, R. (1992). Leaf glucosinolate profiles and their relationship to pest and disease resistance in oilseed rape. *Euphytica,* **63**, 71–83.

Mithen, R., Lewis, B.G., Fenwick, G.R. & Heaney, R.K. (1987). Glucosinolates of wild and cultivated *Brassica* species. *Phytochemistry,* **26**, 1969–73.

Nair, K.S.S. & McEwen, F.L. (1976). Host selection by the adult cabbage maggot, *Hylemya brassicae* (Diptera: Anthomyiidae): effect of glucosinolates and common nutrients on oviposition. *Canadian Entomology,* **108**, 1021–30.

Parkin, I., Magrath, R., Keith, D., Sharpe, A., Mithen, R. & Lydiate, D. (1994). Genetics of aliphatic glucosinolates II. Hydroxylation of alkenyl glucosinolates in *Brassica napus. Heredity,* **72**, 594–8.

Poulton, J.E. & Moller, B.L. (1993). Glucosinolates. In *Methods in Biochemistry vol.9, Enzymes of Secondary Metabolism,* ed P.J. Lea, pp. 209–37. London: Academic Press.

Reed, D.W., Pivnick, K.A. & Underhill, E.W. (1989). Identification of chemical oviposition stimulants for the diamondback moth, *Plutella xylostella,* present in three species of Brassicae. *Entomologia Experimentalis et Applicata,* **53**, 277–86.

Rossiter, J.T., James, D.C. & Atkins, N. (1990). Biosynthesis of 2-hydroxy-3-butenylglucosinolate and 3-butenylglucosinolate in *Brassica napus. Phytochemistry,* **29**, 2509–12.

Rucker, B. & Rudolf, E. (1992). Investigations of the inheritance of the glucosinolate content of seeds of winter oilseed rape *(Brassica napus).* Proceedings of the Eighth International Rapeseed Congress, Saskatoon, Canada, pp. 191–6.

Underhill, E.W. (1980). Glucosinolates. In *Encyclopaedia of Plant Physiology,* eds. E.A.Bell & B.V.Charlwood, vol. 8, pp. 493–511. Berlin, Heidelberg & New York: Springer Verlag.

Underhill, E.W., Wetter, L.R. & Chisholm, M.D. (1973). Biosynthesis of glucosinolates. *Biochemical Society Symposia,* **38**, 303–26.

Index